Acclaim for Jeffrey Rosen's

The Unwanted Gaze

A Washington Post Book World *Favorite Book of 2000*

One of Fortune's *Best Business Books of 2000*

"Rosen explores complex legal cases with a grace and elegance that lawyers and non-lawyers alike should admire. . . . It takes a rare gift . . . to be able to translate the law's arcana for the laity, without oversimplifying. . . . Rosen possesses that rare talent as a writer and thinker. . . . As this groundbreaking book demonstrates, today I have no real privacy. And neither do you."
   —Michael Mello, *The Washington Post Book World*

"Mr. Rosen has given us a measured and bracingly sane examination. . . . His clear voice speaks directly to those of us who still wish to claim a few private moments when Big Brother isn't watching."
   —Francine Prose, *The New York Observer*

"A trenchant essay that speaks not of a distant dictatorship but of a present and ever-more-threatening tyranny, one few are aware of until they become its victims."
   —*The Boston Globe*

"Highly readable. . . . Rosen is both a scholar and a story-teller, a refreshing combination."
—*The San Diego Union-Tribune*

"It is not too extreme to say that everyone in America should read this provocative book."   —*Desert News*

"Read Rosen's book and you'll fully understand the larger historical and legal context for our loss of privacy, and how business is at least as responsible as government for the erosion."   —*Fortune*

"Jeffrey Rosen is America's most penetrating legal journal-ist. . . . There is no work—and no author—I know that does a better job braiding legal scholarship, academic ideas outside law, and popular culture."
—Akhil Reed Amar,
Southmayd Professor of Law,
Yale Law School

"Possibly the most important privacy book in a genera-tion . . . a brilliant exploration of one of today's most pressing social issues."
—Marc Rotenberg, executive director,
Electronic Privacy Information Center

Jeffrey Rosen

The Unwanted Gaze

———————— ✠ ————————

Jeffrey Rosen is an associate professor at the George Washington University Law School and legal affairs editor of *The New Republic*. He is a graduate of Harvard College; Balliol College, Oxford, where he was a Marshall Scholar; and Yale Law School. His essays and book reviews have appeared in many publications, including *The New York Times Magazine* and *The New Yorker*. He lives in Washington, D.C.

The
Unwanted
Gaze

The Unwanted Gaze

The Destruction of Privacy in America

Jeffrey Rosen

With a New Afterword by the Author

Vintage Books

A Division of Random House, Inc.

New York

FIRST VINTAGE BOOKS EDITION, JUNE 2001

Copyright © 2000, 2001 by Jeffrey Rosen

All rights reserved under International and Pan-American Copyright
Conventions. Published in the United States by Vintage Books, a division of
Random House, Inc., New York, and simultaneously in Canada by
Random House of Canada Limited, Toronto. Originally published in hardcover
in the United States by Random House, Inc., New York, in 2000
in slightly different form.

Vintage and colophon are registered trademarks of Random House, Inc.

The Library of Congress has cataloged the Random House edition as follows:
Rosen, Jeffrey.
The unwanted gaze : the destruction of privacy in America / Jeffrey Rosen.
p. cm.
Includes index.
ISBN 0-679-44546-3 (alk. paper)
1. Privacy, Right of—United States. 2. Data protection—Law and legislation—
United States. 3. Computer security—United States. I. Title.
KF1262.R67 2000 342.73'0858—dc21 99-056498

Vintage ISBN: 0-679-76520-4

Book design by J. K. Lambert

www.vintagebooks.com

Printed in the United States of America
10 9 8 7 6 5 4 3 2 1

For Sidney and Estelle Rosen,

my beloved parents

To satisfy a prurient taste, the details of sexual relations are spread broadcast in the columns of the daily papers. To occupy the indolent, column upon column is filled with idle gossip which can only be procured by intrusion upon the domestic circle.

Warren and Brandeis, *The Right to Privacy*

Even the smallest intrusion into private space by the unwanted gaze causes damage, because the injury caused by seeing cannot be measured.

"Hezzek Re'iyyah," Encyclopedia Talmudit

Contents

The
Unwanted
Gaze

Prologue

The Unwanted Gaze

This book began as an effort to understand the constitutional, legal, and political drama that culminated in the impeachment and acquittal of President Bill Clinton. But that
strange and singular confluence of events prompted me to think
about the Clinton impeachment as a window onto a less unusual
phenomenon that affects all Americans: namely, the erosion of
privacy, at home, at work, and in cyberspace, so that intimate
personal information—from diaries, e-mail, and computer files
to records of the books we read and the Web sites we browse—
is increasingly vulnerable to being wrenched out of context and
exposed to the world. What follows is an attempt to explore the
legal, technological, and cultural changes that have undermined
our ability to control how much information about ourselves is
communicated to others. I would also like to consider ways of
reconstructing some of the zones of privacy that law and technology have been allowed to invade.

In January of 1998, when Kenneth Starr began to examine allegations that President Clinton had lied under oath about an

adulterous affair, I became interested in trying to identify the legal forces that converged in Paula Jones's sexual harassment suit and in the subsequent impeachment investigation. Why, for example, were Jones's lawyers permitted to go on a fishing expedition into the President's sexual history, asking him to identify all the women with whom he had sexual relations as governor and president? Merely by accusing Clinton of an unwanted advance, Jones was able to violate not only his privacy but also that of Monica Lewinsky, who was forced to describe her own consensual sexual activities under oath. How could the law permit such unreasonable searches, in which the investigation of the offense seemed more invasive than the offense itself?

The invasions of privacy continued to multiply during the Starr investigation and the impeachment trial that followed. Many examples of the erosion of privacy by means of technology seemed to sit uneasily with the public—the DNA testing, the retrieval of e-mails that Lewinsky and a friend had tried unsuccessfully to delete, the tape recordings, the release of the secret grand jury transcripts on the Internet. But Lewinsky herself was especially unsettled by Starr's decision to subpoena a Washington bookstore for receipts of all of her book purchases since 1995. In her memoir, Lewinsky pointed to the bookstore subpoenas as one of the most invasive moments in the Starr investigation, along with the moment when prosecutors retrieved from her home computer the love letters that she had drafted, but never sent, to the President. "It was such a violation," she complained to her biographer, Andrew Morton. "It seemed that everyone in America had rights except for Monica Lewinsky. I felt like I wasn't a citizen of this country anymore."[1]

Monica Lewinsky is an improbable spokesperson for the virtues of reticence, but her ordeal raises deep questions about recent changes in law and technology that threaten individual control over personal information. In the late eighteenth century, the spectacle of state agents breaking into a suspect's home and rummaging through his or her private diaries was considered the paradigm case of the unreasonable searches and seizures that the framers of the Bill of Rights intended to forbid. In the most famous essay on privacy ever written, published in the *Harvard Law Review* in 1890, Louis D. Brandeis, the future Supreme Court justice, and Samuel D. Warren, his former law partner, announced confidently that "the common law secures to each individual the right of determining, ordinarily, to what extent his thoughts, sentiments, and emotions shall be communicated to others." The legal principle that prevented prosecutors from scrutinizing diaries, letters, books, and private papers, Warren and Brandeis wrote, was the same principle that, in their view, should prevent gossip columnists from writing about the sex lives of citizens. They called that principle the right to an "inviolate personality" and said that it was part of the more general "right to be let alone."

In asserting a right to privacy that could constrain the press, the two lawyers were treading on adventurous ground; yet it was a matter of general agreement, in the 1890s, that the Constitution prohibited prosecutors and civil plaintiffs from rummaging through private papers in search of sexual secrets or anything else. How, then, could that consensus have eroded to the point that Lewinsky's unsent love letters could be retrieved from her home computer? Part of the answer, I will argue, has to do with an unfortunate confluence of decisions by the Supreme Court. The legal forces that culminated in the Clinton impeachment—in particular, the erosion of privacy law, embodied in Fourth and Fifth Amendment protections for individual control over personal information, and the expansion of sexual harassment law, to a point where people can be interrogated about the details of their consensual relationships on the flimsiest of allegations—are the product of surprisingly recent Supreme Court decisions. It was during the 1970s and 1980s, for example, that the principle that private diaries couldn't be subpoenaed as "mere evidence" in civil or white-collar criminal cases was quietly allowed to wither away. And it was during the 1980s and 1990s that the Supreme Court recognized sexually explicit speech and conduct that created a "hostile or offensive working environment" as a form of gender discrimination, a development that made it increasingly difficult for lower courts and employers to distinguish consensual affairs from illegal forms of sexual coercion.

The Lewinsky investigation might never have occurred, however, if these two unfortunate legal trends hadn't converged with a third novel and illiberal law: the Independent Counsel Act, which encouraged a level of inquisitorial zeal in which ordinary prosecutors—constrained, as they are, by time, money, and public accountability—are less likely to indulge. Now that both political parties have experienced the excesses of monomaniacal independent counsels, that law, mercifully, has been allowed to expire. But like a blinding klieg light that exposes the fissures in every surface on which it is turned, the independent counsel law served the jarring yet useful purpose of revealing the fault lines in the legal and technological protections for privacy today.

A hundred years ago, Brandeis and Warren worried that changes in technology as well as law were altering the nature of privacy. "Instantaneous photographs and newspaper enterprise have invaded the sacred precincts of private and domestic life, and numerous mechanical devices threaten to make good the prediction that 'what is whispered in the closet shall be proclaimed from the housetops,' " they wrote. But it was not, in fact, the desire to be let alone that motivated Brandeis and Warren to write their famous article; it was instead the desire to restrict discussion of an intimate family event to the sympathetic boundaries of their own social circle. What outraged Brandeis and Warren was a mild society item in the Boston *Saturday Evening Gazette* that described a lavish breakfast party Warren himself had put on for his daughter's wedding. Although the information itself wasn't inherently salacious, Brandeis and Warren were appalled that a domestic ceremony would be described in a gossip column and discussed by strangers. For this reason, they conceived of privacy in geographic or spatial terms. The press, they wrote, was "overstepping" the "bounds of propriety and of decency," and intruding on "the domestic circle."

At the beginning of the twenty-first century, new technologies of communication have increased the danger that intimate personal information originally disclosed to our friends and colleagues may be exposed to—and misinterpreted by—a less understanding audience. For as thinking and writing increasingly take place in cyberspace, the part of our life that can be monitored and searched has vastly expanded. E-mail, even after it is ostensibly deleted, becomes a permanent record that can be resurrected by employers or prosecutors at any point in the future. On the Internet, every Web site we visit, every store we browse in, every magazine we skim, and the amount of time we spend skimming it, create electronic footprints that increasingly can be traced back to us, revealing detailed patterns about our tastes, preferences, and intimate thoughts. A friend who runs a Web site for political junkies recently sent me the "data trail" statistics that he receives each week. They disclose not only the

Internet addresses of individual browsers who visit his site, clearly identifying their universities and corporate employers, but also the Web sites each user visited previously and the articles he or she downloaded there. Amazon.com has announced a similarly creepy feature that uses zip codes and domain names to identify the most popular books purchased on-line by employees at prominent corporations. (The top choice at Charles Schwab: *Memoirs of a Geisha*.)[2] And under pressure from jittery investors, DoubleClick Inc., the Internet's largest advertising company, had to delay a plan to create elaborate dossiers linking users' on-line and off-line browsing habits with their actual identities.

The sense of violation that Monica Lewinsky and the DoubleClick and amazon.com customers experienced when their reading habits were exposed points to a central value of privacy that I want to explore in this book. Privacy protects us from being misdefined and judged out of context in a world of short attention spans, a world in which information can easily be confused with knowledge. True knowledge of another person is the culmination of a slow process of mutual revelation. It requires the gradual setting aside of social masks, the incremental building of trust, which leads to the exchange of personal disclosures. It cannot be rushed; this is why, after intemperate self-revelation in the heat of passion, one may feel something close to self-betrayal. True knowledge of another person, in all of his or her complexity, can be achieved only with a handful of friends, lovers, or family members. In order to flourish, the intimate relationships on which true knowledge of another person depends need space as well as time: sanctuaries from the gaze of the crowd in which slow mutual self-disclosure is possible.

When intimate personal information circulates among a small

group of people who know us well, its significance can be weighed against other aspects of our personality and character. By contrast, when intimate information is removed from its original context and revealed to strangers, we are vulnerable to being misjudged on the basis of our most embarrassing, and therefore most memorable, tastes and preferences. Monica Lewinsky didn't mind that her friends knew she had given the President a copy of Nicholson Baker's *Vox,* because her friends knew that she was much more than the type of person who would read a book about phone sex. But when our reading habits or private e-mails are exposed to strangers, we may be reduced, in the public eye, to nothing more than the most salacious book we once read or the most vulgar joke we once told. Even if the books we read aren't in any way embarrassing— *Memoirs of a Geisha* is a well-regarded novel—we run the risk of being stereotyped as the *kind* of person who would read a particular book or listen to a particular song. In a world in which citizens are bombarded with information, people form impressions quickly, based on sound bites, and these impressions are likely to oversimplify and misrepresent our complicated and often contradictory characters.

Defenders of transparency argue that more information, rather than less, is the best way to protect us against this sort of misjudgment. We might think differently about a Charles Schwab employee who ordered *Memoirs of a Geisha* from amazon.com if we knew that she also listened to gangsta rap and subscribed to *Popular Mechanics.* But even if we saw the DoubleClick logs of everything she had read and downloaded this week, we wouldn't come close to knowing who she really is. (Instead, we would misjudge her in all sorts of new ways.) In a surrealist world in which complete logs of every citizen's reading habits were available on the Internet, the limits of other citizens' attention spans would guarantee that no one could focus long enough to read someone else's browsing logs from beginning to end. Instead, overwhelmed by information, citizens would change the channel or click to a more interesting Web site. In a world of short attention spans, privacy is necessary to protect citizens from the misjudgments that can result from the exposure of too much information as well as too little information. Filtered or unfiltered, information taken out of context is no substitute for the genuine knowledge that can only emerge slowly over time.

In the same way, gossip that is appropriate to circulate in a closed community of people who know us well can be easily misunderstood when it is retrieved years later and exposed to the world. *The Washington Post,* for example, recently described the case of James Rutt, a man who tried to erase his own past. Rutt had spent a decade unburdening himself in an Internet chat group, and although he was happy to speak candidly in the sympathetic confines of a space that he considered a virtual corner bar, he feared that his musings about sex, politics, and his own weight problem would be used against him in his new position as the chief executive of an Internet company.[3] Fortunately for Rutt, the chat group offered a special software feature called "scribble" that allowed him to erase a decade of his own postings. But as intimate information about our lives is increasingly recorded, archived, and made hard to delete, there is a growing danger that a part of our identity will be confused with the whole of our identity.

William Miller, who has written a book called *The Anatomy of Disgust,* notes that in forming mental pictures, "we often will use the synecdoche, the part for the whole." Rather than remembering everything about a person, in other words, we tend to remember one salient feature. The protagonists of the chapters that follow are individuals who, as a result of invasions of privacy, experienced the phenomenon of the synecdoche. From John Wilkes, the eighteenth-century Whig, whose diaries were seized, to Lawrence Lessig, the Microsoft "special master," whose e-mail joke was misconstrued, to the former dean of the Harvard Divinity School, who was forced to resign after downloading Internet pornography, the subjects of this book found that their public faces were unfairly distorted after snippets of private speech and conduct were exposed to and misinterpreted by an impatient world.

Privacy is necessary to protect all of us from this kind of misinterpretation. A liberal state respects the distinction between public and private speech because it recognizes that the ability to expose in some contexts parts of our identity that we conceal in other contexts is indispensable to freedom. Privacy is necessary for the formation of intimate relationships, allowing us to reveal parts of ourselves to friends, family members, and lovers that we withhold from the rest of the world. It is, therefore, a precondition for friendship, individuality, and even love. In *The Unbearable Lightness of Being,* Milan Kundera describes how the police destroyed an important figure of the Prague Spring by recording his conversations with a friend and then broadcasting them as a radio serial. Reflecting on his novel in an essay on privacy, Kundera writes, "Instantly Prochazka was discredited: because in private, a person says all sorts of things, slurs friends, uses coarse language, acts silly, tells dirty jokes, repeats himself, makes a companion laugh by shocking him with outrageous talk, floats heretical ideas he'd never admit in public, and so forth."[4]

Freedom is impossible in a society that refuses to respect the fact that "we act different in private than in public," Kundera argues, a reality that he calls "the very ground of the life of the individual." By requiring citizens to live in glass houses without curtains, totalitarian societies deny their status as individuals, and "this transformation of a man from subject to object is experienced as shame." Putting a similar point in a different way, the sociologist Erving Goffman argued that individuals, like actors in a theater, need backstage areas where they can let down their public masks, collect themselves, and relieve the tensions that are an inevitable part of public performance.[5] In addition to protecting freedom and self-expression, I will suggest, the privacy of the backstage protects us from the unfairness of being misjudged by strangers who don't have time to put our informal speech and conduct into a broader context.

In America, however, changes in law as well as changes in technology are blurring the boundaries between home and work, reducing the backstage areas in which we can retreat from public view, and increasing the risk that private speech and conduct will be exposed to public scrutiny. Most of us, thankfully, will never have an affair with the president, and those who prove unable to resist this unfortunate temptation in the future will be spared the company of an independent counsel lurking in the galley kitchen. Most Americans, too, will never be deposed in a sexual harassment suit, either as a plaintiff, a defendant, or a witness. Nevertheless, many Americans have their e-mail or Internet browsing habits monitored at work, and one of the most common justifications of employee monitoring offered by courts and management lawyers is the fear of liability for sexual harassment. The Supreme Court's definition of sexual harassment is very vague: it includes not only sexual extortion—this is known as "quid pro quo" harassment: sleep with me or you're fired—but also a more ambiguous category known as "hostile environment" harassment. In 1986, the Supreme Court said that hostile environment harassment includes "unwelcome sexual advances, requests for sexual favors, and other verbal or physical conduct of a sexual nature" that "has the purpose or effect of . . . creating an intimidating, hostile, or offensive working environment." Because it is difficult to know in advance what kind of sexually related behavior or speech a reasonable juror might find hostile or offensive, prudent employers and school administrators, in an effort to avoid liability, have a strong incentive to monitor and punish far more private speech and conduct than the law actually forbids.

The tensions between sexual harassment law and individual privacy should hardly come as a surprise. On the contrary, they are inherent in the legal definition of sexual harassment itself. Harassment law was conceived in the 1970s and 1980s by a group of feminist scholars who insisted that privacy had a darker side, pointing to its capacity to shield from public view the degradation, domination, sexual abuse, and harassment that women may experience in private.[6] Catharine MacKinnon, the most prominent theorist of harassment law, famously encapsulated the feminist critique of privacy in 1989:

> For women the measure of the intimacy has been the measure of the oppression. This is why feminism has had to explode the private. This is why feminism has seen the personal as the political. The private is public for those for whom the personal is political. In this sense, for women there is no private, either normatively or empirically. Feminism confronts the fact that women have no privacy to lose or to guarantee.[7]

The Lewinsky scandal exposed the excesses of the feminist critique of privacy with special force. "This right to privacy is a right of men 'to be let alone' to oppress women one at a time," MacKinnon had argued. "Privacy law keeps some men out of the bedrooms of other men."[8] During the impeachment drama, however, privacy law failed to keep Kenneth Starr out of the bedroom of Monica Lewinsky, whose experience was a dramatic rebuke to the claim that women have no privacy to lose. "Monica felt a brooding sense of outrage that Paula Jones's freedom to sue the President for money should take precedence over her right to privacy,"[9] Andrew Morton reported.

In the wake of the Clinton impeachment, some politicians expressed interest in patching some of the holes in the legal protections for privacy that the Starr investigation helped to expose. But there has been less public interest in examining the excesses in sexual harassment law that also contributed to the debacle. Most Americans are—or claim to be—in favor of privacy; no decent person is in favor of sexual harassment. Moreover, anyone who has joined the workforce during the past generation has benefited from the heightened sensitivity to the perils of sexual harassment, a sensitivity that has helped men and women to work together with greater civility, equality, and mutual respect. The integration of women into the workforce is one of the most important social changes of the postwar era, and the transformation of social norms so that people who inevitably find themselves attracted to one another can interact as professionals is no small achievement. To the extent that sexual harassment law, as it was conceived by its early architects, applied a kind of shock therapy, making clear that any sexual interactions in the workplace may be tainted by unequal power relationships, it deserves credit for improving the lives of American citizens in demonstrable ways.

But it is now hard to ignore the gap between the kind of sexual conduct that most people think should be illegal and the kind the law gives employers and schools an incentive to monitor and to forbid. The Lewinsky affair has challenged us to refine the legal definition of sexual harassment so that the privacy and autonomy of individual women and men are preserved rather than invaded. In thinking further about this challenge, I've become convinced that by focusing on an amorphous vision of privacy that is really a misnomer for the freedom to make intimate decisions about reproduction, the Supreme Court has neglected a more focused vision of privacy that has to do with our ability to control the conditions under which we make different aspects of ourselves accessible to others. In cases like *Roe v. Wade,* where the Supreme Court declared a constitutional right to privacy, the justices were using privacy as a clumsy metaphor for rights of reproductive choice that the Court would later reconceive as an aspect of gender equality.[10] By contrast, in the sexual harassment cases, the justices held that it was a form of gender discrimination to commit certain offenses against personal dignity in the workplace that may be better conceived as invasions of privacy.

Let me be more specific about the kind of privacy I have in mind. In a book called *Behavior in Public Places,* published in 1963, Erving Goffman described the subtle and complicated range of expressions, glances, and gestures that we use to make ourselves accessible to other people in face-to-face interactions, or to maintain boundaries of reserve. Basing his argument on studies of etiquette manuals, Goffman suggested that all of us rely instinctively on a richly calibrated range of "involvement shields" that regulate our encounters with strangers, acquaintances, and intimate friends. In different social settings, we maintain boundaries of reticence that other people are forbidden to cross without mutually negotiated consent. Strangers, for example, expect from each other a zone of privacy that Goffman calls "civil inattention": it's considered rude to stare at strangers whom you encounter in public, although staring at celebrities can be an exception to this rule. At the same time, it may be polite for two strangers to acknowledge each other's presence by briefly glancing at each other from a distance, and then dimming the lights by casting the eyes down as they pass.[11] Anyone who jogs regularly will recognize this ritual: when a runner approaching from the opposite direction fails to look away at the right moment, you may feel either violated or flattered, depending on your point of view, unless both of you are wearing sunglasses, which, as Goffman notes, now play the role in concealing stolen glances that used to be played by parasols and fans.[12]

The rules of civil inattention governing the interaction of men and women have traditionally been even more ornate. In the nineteenth century, some writers on etiquette argued that because a gentleman can never know when a lady wants to be recognized, "[i]t is a mark of high breeding not to speak to a lady in the street, until you perceive that she has noticed you by an inclination of the head."[13] In traditional Muslim societies, any social recognition between the sexes can be interpreted as a prelude to sexual intercourse. Islamic canon law requires women to cover all but their hands and face. Muslim women are expected to look demurely at the ground at the approach of a man, while men are enjoined from gazing directly at women, especially unveiled women.[14] These different examples suggest that although social norms of accessibility vary widely according to culture and context, people have a general expectation that they won't be molested by social overtures to which they haven't explicitly or implicitly given consent.

In America today, sexual harassment codes have arisen as an imperfect substitute for the role etiquette manuals used to play in codifying the social norms that regulate face-to-face interactions. Consider the follow examples of sexual harassment, enumerated by a typical harassment policy at a state university:

> Making sexual innuendoes; turning work discussions to sexual topics; telling sexual jokes or stories; asking about sexual preferences or history; asking personal questions about a person's social or sexual life; making sexual comments about a person's clothing, body, or looks; repeatedly asking out a person who is not interested; telling lies or spreading rumors about a person's sex life. Looking a person up and down (elevator eyes); blocking a person's path; displaying sexual or derogatory comments about men/women on coffee mugs, clothing, and so on; making facial expressions such as winking, throwing kisses; making sexual gestures with hands and/or body movements; giving letters, gifts, and/or materials of a sexual nature; invading a person's body space by standing closer than appropriate or necessary.[15]

Many of the liberties on this remarkable list are better conceived as invasions of privacy than as gender discrimination. "Elevator eyes," for example, in which one person lasciviously looks another up and down, are hard to imagine as an example of gender discrimination. But it is by no means absurd to conceive of leering glances or "elevator eyes" as an invasion of privacy. Indeed, Goffman emphasizes the special difficulty of maintaining civil inattention in small spaces such as railway cars or, as it happens, elevators. He quotes a charming essay from the 1950s by Cornelia Otis Skinner called "Where to Look," which identifies the modern elevator as a particularly trying example of a "where to look situation":

> Any mutual exchange of glances on the part of the occupants would add almost a touch of lewdness to such already over-cozy sardine formation. Some people gaze instead at the back of the operator's neck, others stare trance-like up at those little lights which flash the floors, as if [the] safety of the trip were dependent upon such deep concentration.[16]

To be looked up and down by strangers can be an indignity, although in most cases a minor one. To be looked up and down by your boss can also be an indignity, and in many cases a more serious one. But American law has traditionally been reluctant to punish injuries against individual dignity. "Our system," Brandeis and Warren wrote wistfully, "does not afford a remedy even for mental suffering which results from mere contumely and insult, from an intentional and unwarranted violation of the 'honor' of another."

Other legal systems, whose protections for privacy are less rooted in ideas of private property, have less trouble describing the injury that results when people are observed against their will. Jewish law, for example, has developed a remarkable body of doctrine around the concept of *hezzek re'iyyah,* which means "the injury caused by seeing" or "the injury caused by being seen." This doctrine expands the right of privacy to protect individuals not only from physical intrusions into the home but also from surveillance by a neighbor who is outside the home, peering through a window in a common courtyard.[17] Jewish law protects neighbors not only from unwanted observation, but also from the *possibility* of being observed. Thus, if your neighbor constructs a window that overlooks your home or courtyard, you are entitled to an injunction that not only prohibits your neighbor from observing you but also orders the window to be removed.[18] From its earliest days, Jewish law recognized that it is the uncertainty about whether or not we are being observed that forces us to lead more constricted lives and inhibits us from speaking and acting freely in private places.

Even a fleeting glimpse of intimate activities in a neighbor's house can cause injury, medieval authorities recognized. "To whatever extent the unwanted gaze establishes its sway [over the private domain of another], there is injury, because the damage caused by the gaze has no measure."[19] The injuries that result from a nosy neighbor's unwanted gaze include the danger of being cursed by the evil eye, the fear that gossip and slander will result, and the offense against sexual modesty. But these are manifestations of a broader injury, which is an injury directly to the person, an intrinsic offense against individual dignity. The consensus among medieval jurists, therefore, was that a window overlooking a common courtyard had to be removed even if the individuals whose privacy was violated failed to protest. "One may waive a right only in matters of civil law, where one may give up what belongs to him or may consent to damage to his property; but one is not permitted to 'breach the fences of Israel' or act immodestly so as to cause the Divine Presence (*Shekhinah*) to depart from Israel."[20] Only citizens who respect one another's privacy are themselves dignified with divine respect.

Recognizing unwanted gazes as an offense against privacy and dignity helps us to understand that other forms of sexual harassment commonly experienced by women may be better conceived of as invasions of privacy than as examples of gender discrimination. Anita Allen, a law professor at the University of Pennsylvania who has written an eloquent feminist defense of privacy, argues that leering, sexual insults, catcalls, and offensive touching all constitute a loss of anonymity, which is a form of privacy.[21] But harassment also threatens privacy in the broader sense that I have discussed. By treating a woman as nothing more than a sex object, and allowing a single aspect of her identity—her appearance—to eclipse all other aspects of her identity, an unwanted gaze judges her out of context in the same way that Monica Lewinsky was judged out of context by the exposure of her bookstore receipts. In the 1970s, feminist film critics wrote about the objectifying quality of the "male gaze" in popular movies, which reduced women, in their view, to eroticized body parts fit for ogling by male directors, cameramen, and spectators.[22] These critics are correct, I think, when they describe the indignity that can result from being looked at in a way that substitutes a part of a woman's body for the whole of her personality, but that indignity is more precisely described not primarily as a form of gender discrimination but instead as an invasion of privacy.

Armed with a definition of privacy as a claim about social boundaries that protect us from being simplified and objectified and judged out of context, we can understand that if Bill Clinton exposed himself to Paula Jones in a Little Rock hotel room, the injury she suffered may be better described as an invasion of privacy than as a form of gender discrimination. On public streets and in the workplace, all of us are entitled to expect that we won't be bothered by sexual overtures, unless we've clearly indicated, by our words, gestures, and glances, that the overtures will be reciprocated. Certain spaces—hotel bars, cocktail lounges, even some supermarkets—have traditionally been seen as "open places" where sexual advances are considered permissible behavior rather than an invasion of privacy.[23] The hotel room to which Paula Jones allegedly repaired after allegedly being told by a state trooper that she made Governor Clinton's "knees knock" might be seen in the same light. But even in a hotel room, a sexual advance can invade the recipient's expectations of privacy when it is disproportionate to the kind of attention that she has invited. Paula Jones may have gone to the hotel room prepared to flirt with Clinton, but by all indications she was alarmed and discomfited when, without warning, he allegedly exposed himself to her instead. If Clinton did what Jones accused him of doing, he insulted Jones by invading her privacy.

Similarly, if Clarence Thomas did what Anita Hill accused him of doing—asked her out on dates between five and ten times, discussed in graphic detail the pornographic movies he had seen, joked about pubic hair, exhorted her to watch pornography, and boasted about his sexual prowess—his offenses, too, may be better described as invasions of privacy than as gender discrimination. If her allegations are true, Hill tried to maintain professional boundaries of reserve; Thomas tried to break down those boundaries by treating Hill like a sex object rather than a professional, interjecting the most intimate matters into a space that should be limited to work-related concerns. And Hill, like Jones, suffered from the feelings of humiliation, indignity, and mental distress that are the natural consequence of being misrepresented and judged out of context.

Should the invasions of privacy that Jones and Hill suffered be illegal? I will explore this question further in chapter 4; but my tentative answer is no. Except in the most egregious cases, breaches of our involvement shields have traditionally been punished by social disapproval rather than legal remedies. Law, I will argue, is only one mechanism for deterring and punishing invasions of privacy, and it is often a less effective mechanism than social norms and technological solutions, which can achieve the same purposes without threatening the privacy of innocent people in the process. The law is a blunderbuss rather than a scalpel, and recent scandals in Washington and the workplace have taught us that the effort to provide legal remedies for relatively minor invasions of privacy may inadvertently lead to privacy violations greater than those the law seeks to redress.

"The ability of government, consonant with the Constitution, to shut off discourse solely to protect others from hearing it is . . . dependent upon a showing that substantial privacy interests are being invaded in an essentially intolerable manner," the Supreme Court declared in a case protecting a student's right to wear a jacket declaring FUCK THE DRAFT in a county courthouse. When offensive speech directly and intolerably threatens the privacy of the home, a place with special constitutional protection, the Court has held that the First Amendment rights of the speaker may sometimes yield to the privacy rights of the aggrieved listener. But when the speech in question offends norms of propriety in more public places, courts have generally held that the privacy interests of the listener don't rise to the level of being constitutionally protected and therefore must yield to the First Amendment rights of the speaker, except in the most extreme cases.

It is easy to be misunderstood when writing about this fraught aspect of the broader subject of privacy, and so I should emphasize that my goal is not to eliminate sexual harassment law, but to rethink it in a way that is consistent with the principles of classical liberalism. In particular, I have in mind John Stuart Mill's famous distinction between self-regarding and other-regarding actions. "[T]he individual is not accountable to society for his actions, in so far as these concern the interests of no person but himself," Mill wrote in *On Liberty,* but "for such actions as are prejudicial to the interests of others, the individual is accountable, and may be subjected either to social or to legal punishment, if society is of opinion that the one or the other is requisite for its protection."[24] By this measure, the conduct currently defined as quid pro quo harassment, which is really a form of sexual extortion, clearly should be prohibited by law. Similarly, discriminatory treatment of women based on sexist or stereotypical views of their abilities is a clear example of gender discrimination under the traditional legal tests. But some of the speech and conduct currently embraced by the vague category of hostile environment harassment is a more complicated question. In cases where sexually offensive speech or conduct has no tangible or intangible job consequences, I will argue, invasion of privacy law may be better equipped than discrimination law to distinguish between indignities that are merely embarrassing and those that are serious enough to be illegal.

In arguing that certain sexual advances can invade privacy, I don't mean to deny that they might also be viewed as a form of gender discrimination. My own view, for what it's worth, is that sexual relations between presidents and interns, like sexual relations between professors and their students, are entirely inappropriate, not because they are necessarily coercive, but because the great imbalance of power and experience makes it impossible, in most cases, for the two parties to choose each other on equal terms and to sustain a relationship based on mutual respect. But as Lewinsky's protests show, views about consensual sexual behavior (with whom, how often, what counts as consent) are tremendously varied, and forcing people to justify their own sexual choices in public can be appalling. Throughout this book, I argue that we should preserve private spaces for those activities about which there are legitimately varying views, activities that no one in a civilized society should be forced to submit to public scrutiny. There are areas of sexuality about which there is no legitimate disagreement: no one cares about the privacy claims of child molesters. But the morality of consensual relationships is a complicated question about which reasonable people can disagree. Privacy protects a space for negotiating legitimately different views of the good life, freeing people from the constant burden of justifying their differences.

Instead of presuming to write comprehensively about privacy, I have selected areas that typify but hardly exhaust the subject, and these organize the chapters that follow. Chapter 1, "Privacy at Home," examines the erosion of legal protections for private diaries and letters, and other intimate secrets in the home. Chapter 2, "Privacy at Work," focuses on e-mail privacy, Internet surveillance, and new technologies that allow employers to monitor the reading and writing habits of their employees. Chapter 3, "Jurisprurience," tries to make a case for reconceiving certain forms of hostile environment sexual harassment as invasions of privacy instead. Chapter 4, "Privacy in Court," discusses the injustice of judging someone's character on the basis of isolated misdeeds, an injustice exacerbated by recent changes in the federal rules of evidence. Chapter 5, "Privacy in Cyberspace," discusses the ways that the architecture of the Internet both threatens privacy and offers ways of reconstructing some of the privacy we have lost. And the Epilogue, "What Is Privacy Good For?," examines the social, political, and personal costs of the new threats to privacy, focusing on the relationship between privacy and gossip. Each of these chapters explores different aspects of the same phenomenon: the danger of misjudging people by confusing information with knowledge in an economy that is increasingly based on the recording and exchange of personal information.

It is surprising how recently changes in law and technology have been permitted to undermine sanctuaries of privacy that Americans took for granted throughout most of our history. But even more surprising has been our tepid response to the increasing surveillance of our personal and private life. Just as citizens in the Soviet Union responded to constant surveillance by having private conversations in parks and making telephone calls from subways, Americans have been similarly passive in adjusting our lives to the intrusions of technology, by avoiding the use of e-mail for private communications, for example, and paying with cash to evade detection by direct marketers. There is nothing inevitable about the erosion of privacy, just as there is nothing inevitable about its reconstruction. We have the ability to rebuild the private spaces we have lost. But do we have the will?

Chapter 1

Privacy at Home

M*onica's Story* is not the most rarefied of sexual memoirs, but it achieves a raw eloquence when describing its heroine's fury at the invasions of her privacy by the "moon-faced figure of Kenneth Starr," whom Lewinsky considers "almost a personification of the 'Big Brother' of Orwell's future." For Lewinsky, we learn, one of the most degrading moments in the Starr investigation was the moment that prosecutors confronted her with drafts of love letters that she had written to the President but never sent. The drafts had been exhumed from her home computer, which prosecutors obtained by issuing a subpoena. Lewinsky's biographer, Andrew Morton, describes her reaction when she was presented with her unpublished thoughts. "Not content with snatching her body, Starr's deputies were now invading her mind. They had exposed her sex life and dissected her personality; now they wanted to scrutinize her very soul. It was an intrusion too far." Lewinsky started to cry, and then screamed at Karen Immergut, the deputy independent counsel, "This is so *wrong*! Do you not understand that no one else is supposed to read this?"[1] Immergut promised her that the letters would be placed under lock and key, and indicated that they might not appear in Starr's report to Congress. But they did appear, and Congress then released them to the world.

One of the virtues of the Starr investigation was to remind Americans how little our legal system cares about privacy today and how much more robustly intimate secrets were protected in the not-so-distant past. The subpoenas issued by Starr were perfectly legal, but for most of American history, many of them would have been suppressed as clear violations of the Fourth Amendment to the Constitution, which declares that "the right of the people to be secure in their persons, houses, papers, and effects, against unreasonable searches and seizures, shall not be violated." This chapter will explore the reasons that constitutional protections for private papers and diaries evaporated during the past few decades, and will suggest ways that they might be resurrected to restrain state agents from invading privacy while preserving the freedom of the press.

To appreciate how dramatically privacy protections have eroded, it's useful to compare the stories of two coltish legislators—John Wilkes, an eighteenth-century Englishman, and Bob Packwood, a twentieth-century American—who tried to conceal their personal diaries from overreaching prosecutors, with very different results. Wilkes is largely forgotten today, but his suit against King George's minions for breaking into his London house, in 1763, and seizing his private papers was so galvanizing for the American revolutionaries that the Sons of Liberty in Boston—a group that included John Adams and John Hancock—insisted that "the fate of Wilkes and America must stand or fall together."[2] Wilkes had been elected to Parliament in 1757. Although Samuel Johnson considered him to be one of the ugliest men he had ever met, Wilkes boasted that it took him "only half an hour to talk away my face," and he was known as a highly successful womanizer. He belonged to a secret society known as the Hell-Fire Club, whose members wore white robes and masks and held delirious orgies in a secret chapel adorned by an enormous sculpture of a phallus. But Wilkes got into trouble by founding a Whiggish scandal sheet called *The North Briton*, a kind of eighteenth-century Drudge Report, whose politics are a little hard to reconstruct today. Several issues were devoted to accusing Lord Bute, the secretary of state, of having an affair with George III's mother.

The king, however, was especially offended by *North Briton* No. 45, a violent attack on a speech of his praising an obscure German peace treaty signed by Bute. Lord Halifax, the Ken Starr figure, issued a general warrant authorizing the arrest of the printers, publishers, and authors of *North Briton* 45, without identifying them by name. After stopping by his printer's office, where he apparently destroyed the original manuscript, Wilkes returned to his house, where he was arrested and transported to the Tower. The king's minions then broke into his house, forced open the drawers of his writing desk, and seized his diaries and private papers. At a hearing before Judge Pratt, later celebrated by American colonists as the heroic Lord Camden, Wilkes objected that "my house [has been] ransacked and plundered; my most private and secret concerns divulged. . . . Such inhuman principles of star-chamber tyranny will, I trust, by this court . . . be finally extirpated." After concluding that Wilkes was immune from arrest as a member of Parliament, Judge Pratt set him free.

Soon after Wilkes's release, an anonymous pamphlet written about his case summarized the reasons that private papers should be immune from seizure in civil and criminal cases. "Papers," the author wrote, "are our closest confidants; the most intimate companions of our bosom; and next to the recesses of our own breasts, they are the most hidden repository we can have." If private papers could be seized, and our confidential communications with others could be exposed, people would take care not to record their confidences in the first place, and friendship, intimate disclosures, and even freedom of thought would be inhibited. "[H]e will be the wisest man that corresponds the least with others, and the most prudent who writes very little, and keeps as few papers as he can by him. None but a fool in this case will have any secrets at all in his possession."[3]

Anticipating Monica Lewinsky's complaint to her prosecutors, the pamphlet also objected to the notion that Wilkes's unpublished thoughts could be used as evidence against him in a future criminal proceeding: "A man's WRITINGS lying in his closet, NOT PUBLISHED, are no more than his thoughts, hardly brought forth even in his own account, and, to all the rest of the world, the same as if they yet remained in embrio in his breast. . . ."[4] The author concluded by comparing the investigation of Wilkes to that of Algernon Sydney, the seventeenth-century poet who was executed for treason on the basis of an unpublished essay found in his home.

Wilkes sued the king's messengers for trespassing on his property, and he urged a number of other publishers and printers who had been arrested under general warrants to do the same. A jury awarded Wilkes one thousand pounds in damages—a ruinous amount in its day. In a related case, John Entick, the author of allegedly libelous articles in a paper called the *Monitor,* sued Nathan Carrington and three other messengers for trespass after they forced their way into his house, broke open his cabinets, and rummaged through his books and papers, seizing several hundred pamphlets and charts. In his landmark opinion siding with Entick, Lord Camden noted three reasons that English law didn't allow the government to search for or seize private papers, with or without a valid warrant. First, he noted, the offense of libel, which was essentially a thought crime, didn't justify seizing private papers. Simply possessing a libelous document was illegal, and therefore any citizen's home would be potentially vulnerable to an intrusive search if libelous writings could be seized. Because it would be impossible for officials to distinguish innocent from illegal material without reading all the papers they found, "the secret cabinets and bureaus of every subject in this kingdom will be thrown open to the search and inspection of a messenger." Second, Lord Camden noted that English law didn't permit the government to seize papers for use as "mere evidence" in criminal cases. According to the theories of the day, government officials could seize contraband or other illegal goods, which a suspect had no right to possess in the first place, but mere evidence of guilt, such as letters or diaries, were the suspect's "dearest property," and therefore immune from search. Finally, Camden agreed that "paper searches" were impermissible because they amounted to a kind of compelled self-incrimination. As he put it in the Wilkes case, "Nothing can be more unjust in itself, than that the proof of a man's guilt shall be extracted from his own bosom."[5]

Wilkes's victory in court didn't go over well with his prudish colleagues in Parliament. After his vindication, Wilkes made the mistake of publishing a pornographic parody of Alexander Pope's *An Essay on Man,* entitled *An Essay on Women,* and in a turn that foreshadowed the Packwood case, the House of Lords condemned it as a "most scandalous, obscene and impious libel."

The House of Commons then voted to expel Wilkes for the "false, scandalous, and seditious libel" of writing *North Briton* 45. After hiding out in Paris, Wilkes returned to England to accept his fate. Finally released from Kings Bench Prison in 1770, Wilkes was elected Lord Mayor of London, and giddy American colonists named towns and infants—from Wilkes-Barre, Pennsylvania, to John Wilkes Booth—in his honor.

More than two centuries later, when Bob Packwood, Republican from Oregon, tried to conceal his diaries from his fellow legislators, he found that the legal protections for private papers had evaporated. Packwood served as a senator for twenty-six years, from 1969 to 1995. More than two dozen former female employees and lobbyists accused him of making unwanted sexual advances during this time, but most of them described incidents that had occurred before the Supreme Court decision, in 1986, that definitely recognized sexual harassment as a civil offense. The exception was Packwood's former press secretary, who claimed that in 1990, after an evening of drinking with colleagues, the Senator had tried to kiss her good night but had retreated after being rebuffed. Summoned before the Senate Ethics Committee, Packwood tried to argue that the advance wasn't unwanted, suggesting that a year after the incident, while drinking wine and listening to Frank Sinatra in his office, the press secretary had been so bold as to kiss him. On being asked to corroborate this claim, Packwood confessed that he had written about it in his diary.

This gave the Ethics Committee an opening to subpoena all of the diaries that Packwood had dictated between 1989 and 1993. Alarmed at the thought of his colleagues rummaging freely through his most private thoughts, Packwood suggested that the committee should instead appoint an "independent examiner" to review the diaries and decide which passages were relevant to the charges of sexual misconduct. (The former appellate judge whom Packwood suggested to conduct the review was, as it happened, Kenneth Starr.) When the committee rejected this request, Packwood grew despondent. "I felt like the very privacy of my heart was being ripped out," he told me later.

Packwood's lawyers then contested the subpoenas, citing a famous opinion from 1886, *Boyd v. U.S.*, in which the Supreme Court had recited the story of John Wilkes, and then announced that subpoenaing a defendant's private business papers in order to use them against him was both an unreasonable search and a form of compelled self-incrimination, violating both the Fourth and Fifth amendments, which "run almost into each other." The Boyd opinion, which Justice Louis D. Brandeis later praised as "a case that will be remembered as long as civil liberty lives in the United States,"[6] was, arguably, an odd occasion for the Court to strike a blow for the sanctity of private papers: it involved a subpoena not for personal diaries but for a business invoice, which the government alleged would show that a Philadelphia company had imported glass without paying the necessary customs duties. But in an expansive opinion, Justice Joseph P. Bradley struck down the federal statute authorizing the subpoena, quoting approvingly from Lord Camden's judgment in *Entick v. Carrington*. The prohibitions on searches of private papers expressed in Camden's opinion, Bradley held, "apply to all invasions, on the part of the Government and its employees, of the sanctity of a man's home and the privacies of life. It is not the breaking of his doors and the rummaging of his drawers that constitutes the essence of the offense; but it is the invasion of his indefeasible right of personal security, personal liberty and private property." In a stirring conclusion, Justice Bradley announced that "any compulsory discovery by extorting the party's oath, or compelling the production of his private books and papers, to convict him of crime or to forfeit his property, is contrary to the principles of a free government. It is abhorrent to the instincts . . . of an American."[7]

Unmoved by Packwood's citation of the Boyd case, Judge Thomas Penfield Jackson of the U.S. District Court in Washington, D.C., ordered him to turn over his diaries to the Senate in 1994. The nineteenth-century right to privacy, Jackson noted in a remarkably breezy opinion, had been chipped away by subsequent Supreme Court decisions that were initially motivated by a single purpose: eradicating white-collar crime. In the years leading up to the Progressive era, it became clear that if people could refuse to turn over their corporate records in response to grand jury subpoenas, then it would be impossible to enforce antitrust laws or railroad laws, and the regulatory state would come to a halt. Well before the New Deal, the Court decided that the only way to investigate corporate crime would be to give prosecutors broad power to subpoena witnesses and to produce documents. And in 1948 the New Deal Court held that the Fifth Amendment wasn't violated by requiring someone to produce records that the government had ordered him to keep, no matter how incriminating or embarrassing the records might be.

But the Warren and Burger Courts went further still, delivering the coup de grâce for constitutional privacy protections. In the sexual privacy cases leading up to *Roe v. Wade*, the Court waxed grandiloquent about "the sacred precincts of the marital bedroom." But the right to privacy in these cases turned out to be a confusing metaphor for the very different right to make personal decisions about procreation. Meanwhile, in a series of less familiar criminal procedure cases, the Court dramatically expanded the power of the police to conduct intrusive searches and, in the process, threatened the ability of innocent people to control the disclosure of personal information. Judges have a natural tendency to favor the state when balancing the interests of prosecutors against the interests of criminals, and any society that ties its privacy to the rights of the accused is a society in which the legal protections for privacy will quickly evaporate.

That, in any event, is what happened to privacy in America. In the 1960s, as part of an effort to constrain violent and racist police forces in the South, the Warren Court held that evidence obtained in unconstitutional searches had to be excluded from criminal trials. But the justices soon realized that they'd painted themselves into a corner: if every search, no matter how trivial, had to meet elaborate constitutional standards, including a warrant issued by a magistrate who has probable cause to believe a crime has been committed, law enforcement would be impossible. The Court's hastily improvised solution was to pretend that all sorts of dramatic intrusions on privacy, such as planting bugs in people's clothing, rummaging through their trash, and spying on them with high-powered binoculars, weren't really searches or seizures in the first place. The result was a legal climate that constricted the constitutional protections for privacy at the very moment that techniques of surveillance were growing more invasive.

Meanwhile, even as the Court was narrowing the definition of impermissible searches, it was expanding the scope of permissible warrants and subpoenas. In the eighteenth and nineteenth centuries, as I mentioned, when vigorous protections for private property prevailed, magistrates could not issue warrants for "mere evidence" that a crime might have been committed. "To enter a man's house by virtue of a nameless warrant, in order to produce evidence, is worse than the Spanish Inquisition," Lord Camden thundered in 1763.[8] Only the fruits and instrumentalities of a crime, such as contraband, could be seized, on the theory that they didn't belong to the suspect in the first place. In 1967, however, Justice William Brennan, of all people, issued one of the most pro-law-enforcement opinions in American history: he abandoned the ancient distinction between seizures of evidence and the fruits of a crime and held that government, armed with a warrant, could search and seize "mere evidence" as well as contraband. In dissent, William O. Douglas—the same justice who, in a famous case striking down restrictions on the use of contraceptives, discovered a right to sexual privacy in the "penumbras, formed by emanations" from the Bill of Rights—objected that Brennan's opinion threatened "the choice of the individual to disclose or to reveal what he believes, what he thinks, what he possesses."[9]

Until the Packwood affair, some courts were still trying to maintain the embattled distinction between business papers and private diaries, which the Supreme Court has never explicitly repudiated. (When Joshua Steiner, the former Treasury aide, voluntarily turned over his diary to congressional investigators in 1994, his lawyer advised him there was a chance he might be able to fight the subpoena.) But in the wake of the Packwood decision, citizens can no longer reasonably expect that their diaries will be considered any more private than their financial records. The consequences are precisely those predicted by Wilkes's supporters: in a world where government officials must assume that their most intimate recorded thoughts may be exposed, they are increasingly taking care not to record their thoughts in the first place.

In a hearing before Judge Norma Holloway Johnson, Robert

Bittman, a deputy independent counsel, defended Kenneth Starr's decision to subpoena Monica Lewinsky's bookstore purchases, noting that federal prosecutors in the Unabomber and Oklahoma City bombing cases had also examined the suspects' private diaries. "She has not been charged with anything," one of Lewinsky's lawyers countered, and therefore "she has the right to have her privacy protected." In light of recent case law, however, Lewinsky's lawyer couldn't make the stronger argument that, even if Lewinsky were to be accused of perjury, her privacy should have been respected rather than invaded. There is, after all, a world of difference between committing mass murder and lying about a consensual affair, and any sane legal system that prohibits "unreasonable" searches and seizures should at least try to distinguish between the two crimes in deciding whether or not a witness's reading habits or intimate thoughts should be exposed to public view.

That, in the end, is the lesson of John Wilkes, who may have been guilty of seditious libel, but was fortunate enough to live in an age in which judges and juries didn't think an accusation of seditious libel was a serious enough crime to justify rummaging through a man's private papers and divulging his sexual secrets. Indeed, Wilkes's supporters argued explicitly that the benefits of punishing seditious libel should be weighed against the danger of politically motivated prosecutions and the accompanying danger of intrusive inquiries into the private thoughts of innocent citizens. It is a foolish system of criminal justice that can't tell the difference between murder and adultery. When it comes to physical strip searches, courts today have no difficulty recognizing that invasions of privacy that might be reasonable in the investigation of serious crimes can be unreasonable in the investigation of less serious crimes. From 1952 until 1979, for example, police in Chicago routinely strip-searched female prisoners whom they had arrested for minor traffic violations.[10] Happily, times change: by 1986, the U.S. Court of Appeals in New York had no hesitation in concluding that it was unconstitutional for the police to subject a woman to a strip search after they arrested her for the misdemeanor of filing a false crime report.[11] But when confronted with mental strip searches, judges have relinquished the tools for distinguishing between violent crimes and thought crimes. The law no longer encourages them, as it should, to balance the intrusiveness of the search against the seriousness of the offense.

To restore this balancing test, Congress might consider listing the crimes that are serious enough to justify the search of private papers, although congressional laundry lists are hardly insulated from political pressure. In 1968, for example, Congress recognized that wiretapping posed such a serious threat to privacy that it could be justified only for especially serious crimes, such as espionage, treason, and crimes of violence. But although wiretapping was authorized for only twenty-six crimes in 1968, there were ninety-five crimes on the list in 1996. In that year, 71 percent of all the wiretaps authorized involved drug cases rather than crimes against the state.[12] Another alternative might be for Congress to create new legal institutions for protecting privacy. Perhaps special grand juries could be empaneled to evaluate the reasonableness of subpoenas and warrants, balancing the intrusiveness of the search against the seriousness of the crime.

Subpoenas are ordinarily considered less threatening to privacy than warrants, because they allow the recipient to surrender the specified items, rather than permitting an officer of the state to rummage freely through a home or office. But a broad subpoena that allows prosecutors to retrieve all the data on a suspect's hard drive looks uncomfortably like a general warrant, which authorizes an unconstrained fishing expedition without specifying the places to be searched or the things to be seized. The fact that private information on computers is hard to delete—in her grand jury testimony, Monica Lewinsky confessed that she had tried unsuccessfully to erase her private e-mails at home, without realizing that prosecutors could retrieve them—makes the threat to privacy in cases involving computer searches all the more acute.

How should the Fourth Amendment be interpreted in a computer age, to protect the privacy that the eighteenth-century framers had in mind when they prohibited general warrants? Arguably, the courts should require some kind of filtering mechanism to prevent prosecutors from riffling through a mass of innocent documents in search of potentially incriminating ones, even with a warrant or subpoena. Instead of allowing Starr to scrutinize Lewinsky's computer, for example, Judge Norma Holloway Johnson could have insisted on reviewing the files herself, and disclosed to the prosecutors only material that was clearly relevant to their investigation and didn't unreasonably threaten Lewinsky's privacy. Or, if Judge Johnson didn't feel that she had the time to undertake such an extensive review, she could have appointed a special privacy master to play the role that Bob Packwood had begged Congress to assign to Kenneth Starr during the investigation of Packwood's diaries, sifting through the hard drive and separating relevant from irrelevant material.

If technology poses new threats to privacy, it also offers new ways of reconstructing some of the privacy we have lost. An essay in the *Yale Law Journal* recently imagined an ingenious computer program called "the worm" that the government can dispatch to enter your computer without notice and scan your hard drive for specified words or images.[13] If the worm finds what it is looking for, it can alert the FBI; if not, it destroys itself, leaving no trace of its presence. Is the worm consistent with the values of the Fourth Amendment? In some respects, it looks precisely like the general warrants that the framers of the Fourth Amendment meant to prohibit: it can enter thousands of computers without probable cause to believe that a crime has been committed, and it searches broadly without particularized suspicion of people or places. In other respects, the worm avoids all of the spillover effects that led the framers of the Fourth Amendment to condemn general warrants: rather than exposing innocent as well as illegal material to public view, it focuses on the illegal material with greater precision.[14]

In a privacy utopia in which police were devoted to identifying the guilty and protecting the innocent, Arnold Loewy of the University of North Carolina has written, each police officer would be equipped with an "evidence-detecting divining rod" that would alert to evidence of crime in the most efficient manner possible, singling out the house, the room, the drawer, and the guilty evidence itself, without exposing evidence of innocent activity in the process.[15] Like a marijuana-sniffing dog or an airport metal detector, the computer worm has some of the virtues of the divining rod, protecting the privacy of the innocent while exposing the sins of the guilty. And courts sensitive to privacy might think about creative variations on the focused search embodied by the worm. They might require prosecutors to submit for judicial approval a list of particular words or phrases that they hoped to find on, for example, Monica Lewinsky's hard drive—"North Gate," "Betty Currie," or perhaps "thong"— rather than permitting them to trawl through Lewinsky's entire computer.

In civil cases involving the seizure of computer hard drives, in which innocent and potentially incriminating documents are hopelessly intermingled, some courts have suggested that the officers should hold the computers until a magistrate specifies the conditions under which they may be searched.[16] When large quantities of information are seized, these courts have suggested, the officers should apply for a second warrant, to ensure that the search will be focused only on relevant documents.[17] By ensuring that a neutral magistrate carefully monitors the scope of computer searches, this approach avoids the dangers of general rummaging through private papers that the framers of the Fourth Amendment were determined to prohibit.

I've argued that we need to think creatively about resurrecting the constitutional restraints on government searches of private papers that used to be taken for granted. But legal restrictions on invasions of privacy by the press and by private citizens are a different matter. The most serious violation of Monica Lewinsky's privacy wasn't simply the search of her computer files by Kenneth Starr; it was Starr's decision to include her grand jury testimony, including her unpublished love letters, in his report to Congress, and Congress's subsequent decision to publish the Starr report on the Internet. Similarly, the cruelest moment in the Bob Packwood drama wasn't the congressional subpoena; it was the gratuitously mean-spirited decision, by *The Washington Post,* to publish excerpts from Packwood's diaries in its gossip column, "The Reliable Source." The excerpts had nothing to do with the sexual harassment charges, but instead held up for ridicule Packwood's private musings on his favorite recipe for baked apples, his favorite Washington supermarket, and his fondness for the music of Edvard Grieg.

When Dan Burton, the head of the House Committee on Government Reform and Oversight, was accused in 1998 of violating the privacy of former associate attorney general Webster Hubbell by releasing edited transcripts of his telephone calls from federal prison, Burton's response was to distribute unedited copies of the tapes to all those who asked for them. Soon afterward, in preparation for an article about the violation of Hubbell's privacy, I found myself sitting at home and eavesdropping on Hubbell and his wife as Mrs. Hubbell discussed her plans for dinner. She remarked that the children wanted meat loaf. Hubbell gently reminded his wife that meat loaf wasn't his favorite food. "Have it before I get home," he said. "I just don't like meat loaf, OK?" His wife clucked indulgently. "Poor Webby," she sighed.

What rankles about the exposure of Packwood's food preferences—or Webster Hubbell's—isn't that the information itself is especially illicit. It is, instead, the breaching of boundaries: information that might be appropriate to share with friends or acquaintances has been taken out of context and exposed to the world. The Packwood, Hubbell, and Lewinsky cases remind us that the press, when it publishes private facts that are revealed in legal proceedings, is often responsible for invasions of privacy far more dramatic than those precipitated by the subpoenas or warrants that extracted the information in the first place.

In a society that respects the freedom of the press, however, the solution to these privacy invasions isn't to prohibit journalists from publishing information they have obtained from police or prosecutors. Instead, courts should prohibit police or prosecutors from sharing certain private information with the press. Some courts have recognized that the reasonableness of a search or seizure may turn on how widely the police publicize the information after they have obtained it. In a case arising out of the investigation of Lyle and Erik Menendez, who were convicted of murdering their parents in Beverly Hills, law enforcement officials searched the home of the brothers' psychotherapist and made a videotape of the search. The warrant authorizing the search said nothing about a videotape, but the camera captured a glimpse of the psychotherapist's wife in a bathrobe, as well as of files revealing the names of other patients. A trial judge initially ordered the video to be released to the press, on the grounds that it was a public record. In a powerful opinion, however, a California appellate court disagreed. "It is one thing to be forced to submit to a search of one's home under color of warrant; it is quite another matter to be forced to have the whole world accompany the master during his search by watching a videotape showing everything the master did and saw during the search," the court held. "Public disclosure in this case implicates not only [the psychotherapist's] privacy interests, but his Fourth Amendment rights."[18]

More recently, in two cases decided in 1999, the Supreme Court agreed that being forced to disrobe before the world is a far more serious invasion of privacy than being forced to disrobe before a single officer. In one case, CNN had mounted hidden video cameras and microphones on government cars and agents, and secretly recorded a suspect after the agents theatrically forced their way into his house.[19] In the other case, a *Washington Post* photographer followed the police into a suspect's house and snapped pictures of his parents in their underwear.[20] All nine justices agreed that the Constitution prohibits federal and state officials from bringing members of the media into a suspect's home, unless the presence of third parties is necessary to execute a warrant.

The Fourth Amendment prohibits searches and seizures conducted by government agents. But should citizens be able to sue the press for invasions of privacy when it operates independently of the state? This was the focus of Brandeis and Warren's famous essay in 1890, but their answer was not entirely satisfying. In trying to identify a legal principle that might constrain the tabloid press from publishing gossip about Warren's daughter's wedding, Brandeis and Warren began by identifying the injury that resulted from the publication of true but intimate private facts. Slander and libel law protect the value of your reputation against inaccurate portrayals by others and provide a remedy for the damage "done to the individual in his external relations to the community by lowering him in the estimation of his fellows." By contrast, as Brandeis and Warren touchingly put it, invasions of privacy "affect a person's estimate of himself and his own feelings." Here Brandeis and Warren faced a dilemma. American law had traditionally protected privacy under the guise of protecting private property. The Fourth Amendment is founded on the theory that a person's "houses, papers, and effects" are his dearest property, which can be invaded only by physical trespass. Even slander and libel law could be conceived as an injury to property, if one thinks of all citizens as having a property interest in their reputations. But the injuries caused by invasions of privacy are "spiritual" rather than "material," Brandeis and Warren noted, and are best described as the "intentional and unwarranted violation of the 'honor' of another."

So Brandeis and Warren set out to propose a new legal theory for protecting the spiritual interests threatened by journalists and private citizens who invade privacy. They focused on the existing legal protection for casual letters and diary entries. If "a man records in a letter to his son, or in his diary, that he did not dine with his wife on a certain day," they noted, "no one into whose hands those papers fall could publish them to the world, even if possession of the documents had been obtained lawfully." Nor could anyone publish a list summarizing the content of the letters. But under the guise of protecting private property in letters, they argued, English courts were really protecting a different interest: "It is not the intellectual product, but the domestic occurrence," they noted, that deserved to be shielded from public view. The same principle of freedom from unwanted observation, they said, should be extended "to protect the privacy of the individual from invasion either by the too-enterprising press, the photographer, or the possessor of any other modern device for recording or reproducing scenes or sounds." The right to privacy, the Bostonians ringingly concluded, "implies the right not merely to prevent inaccurate portrayal of private life, but to prevent its being depicted at all."

Brandeis and Warren argued convincingly that the press can precipitate violations of honor when it wrenches personal information out of context. But when the Boston lawyers asserted the ability to control the conditions of our own exposure as a legal right, they raised troubling questions, not only of freedom of the press, but also of the freedom of nosy neighbors. Unless we pull down the curtains and never leave the house, none of us can avoid being observed, and therefore judged, fairly and unfairly, by others. The idea that the law can give us complete control over the conditions of our self-presentation proved, in the century since Brandeis and Warren's article was published, to be idealistic, and in some cases dangerous to free expression.

In the twentieth century, the common law of torts followed Brandeis and Warren's suggestion by recognizing four distinct categories of invasion of privacy. The first is "appropriation of another's name or likeness,"[21] and the familiar example is an advertising campaign that uses a celebrity's name without permission. Second, there is "publicity that unreasonably places another in a false light before the public,"[22] which occurs when a person's picture is used to illustrate a book or article with which he has no real connection. Third, there is unreasonable intrusion upon the solitude or seclusion of another. The classic case, decided nine years before Brandeis and Warren's article was published, involved a young man who spied on a woman during childbirth by impersonating her doctor; more recent cases involve landlords or homeowners who hide microphones or video cameras in their tenants' bedrooms or their nanny's bath.[23] Finally, there is the public disclosure of private facts, or giving "publicity to a matter concerning the private life of another." In order to be actionable, the matter publicized has to be "highly offensive to a reasonable person" and "not of legitimate concern to the public."[24] Defendants in these cases range from a creditor who put up a sign in the window of his garage announcing that a debtor wouldn't pay the money he owed, to a moviemaker who humiliated a respectable socialite by making a movie about her former life as a prostitute and murder suspect, in which he identified her by name.

The "intrusion on seclusion" and "publication of private facts" torts are potentially important tools for remedying outrageous invasions of privacy by employers or fellow employees. But the requirement that the exposure in question has to be both highly offensive and not of legitimate public concern has made the invasion of privacy torts hard to enforce against the press. In 1937, for example, *The New Yorker* published a "Where Are They Now?" article by James Thurber that included intimate details about the personal life of a math prodigy who had graduated from Harvard at sixteen, then fizzled out, and was languishing as an eccentric recluse. The man sued and lost, on the grounds that former child prodigies are inherently newsworthy. Then, in 1993, an African-American man named Luther Haynes sued Nicholas Lemann, the author of a book about the migration of blacks from the rural South to the North entitled *The Promised Land,* for describing his drinking, philandering, and neglect of his children. He lost, too. The social history of blacks in the inner city, Judge Richard Posner held for the U.S. Court of Appeals in Chicago, is a subject of "transcendent public significance."[25]

Brandeis and Warren recognized that public figures, whom they defined narrowly as aspiring officeholders, can expect less privacy than other citizens, because the press in a democracy must be free to discuss matters of public concern. Expanding the category of public figures far beyond Brandeis and Warren's narrow group of aspiring officeholders, courts today tend to hold that even the most private details of the sexual identity of public figures can be exposed without consent, on the grounds that the disclosures are inherently newsworthy.

Consider the public "outing" of someone who prefers to keep his or her sexual orientation private. This is an assault on a central aspect of privacy, which is the ability to control the face we present to the world. In 1997, David Brock, the conservative writer, apologized to President Clinton for publishing in *The American Spectator* the article about Paula Jones that provoked her lawsuit. In trying to explain Brock's change of heart, Jones's spokeswoman, Susan Carpenter-McMillan, suggested on television that Brock was having a love affair with Hillary Clinton's former press secretary, Neel Lattimore. "I was sitting in my office looking up and there was Susan Carpenter-McMillan saying on TV that I was the love of David Brock's life and we were soul mates and had been living together," Lattimore says. "It was completely ridiculous—we're just friends—but all of a sudden it ended up on *Meet the Press*."

Although Lattimore had never made a point of concealing his sexual orientation from friends and colleagues, he had chosen to keep his personal life private in his public appearances. Suddenly hearing it discussed on network TV was an "out-of-body experience," he says. "You want to breathe but you can't think about how to bring air back in your lungs again. . . . It's like watching a storm destroy your house when you're not in it."

Lattimore considered suing for invasion of privacy, but then thought better of it. Courts have been generally unsympathetic to suits by public figures who have been outed, reasoning with exquisite circularity that the sexual orientation of celebrities is newsworthy because readers tend to be interested in it. In 1975, when an ex-marine named Oliver Sipple saved President Gerald Ford's life by striking the arm of his assailant, Sara Jane Moore, just before she fired her gun, the *San Francisco Chronicle* published an article revealing that Sipple was prominent in the gay community. Sipple sued for invasion of privacy, but a California court dismissed the suit, holding that the article had been prompted not by "a morbid and sensational prying into [Sipple's] private life" but rather by "legitimate political considerations, i.e., to dispel the false public opinion that gays were timid, weak and unheroic figures."[26]

The court's logic—asking how any man can object to being portrayed as manly—seems presumptuous as well as cruel: Sipple hadn't chosen to cast himself as a role model. And sexual identity is the most personal of all aspects of identity. Who can say whether a reasonable person in San Francisco would object to being portrayed as gay? Sipple, for his part, objected strenuously: the involuntary disclosure of his sexual orientation led to his estrangement from his family and to psychological distress. Ten years after he saved the President's life, Sipple committed suicide.[27]

The brutal outing of Oliver Sipple shows the central value of privacy in allowing us to reveal in some contexts parts of our identity that are withheld in others. Oscar Wilde, too, felt justified in denying his homosexuality to his lawyer, although not to his close friends, because he felt that being labeled a sodomite in a society that considered homosexuals to be criminals would misrepresent him and fail to do justice to his own self-conception as a lover of beauty in its highest form.[28] But despite the tragic personal consequences that often result from the disclosure of true but embarrassing private facts, it's appropriate, in a country that takes the First Amendment seriously, that invasion of privacy suits against the press rarely succeed, except in truly lurid cases of hidden camera investigations that transgress the bounds of civilized behavior.[29] Although the "newsworthiness" defense has steadily expanded to the point where it is "so overpowering as virtually to swallow the tort,"[30] as a leading First Amendment scholar put it, perhaps circularity is inherent in any "newsworthiness" standard—unless judges are to decide on their own what people *should* be interested in.

Is there nothing left to the branch of privacy law that prohibits the publication of private facts? Consider the recent epidemic of sexual memoirs, exemplified by Joyce Maynard's *At Home in the World* and a stream of less-known effusions. Many of these books describe intimate encounters that are mortifying for all concerned. A sexual memoir is the equivalent of a literary strip search, depriving the unsuspecting partner of the most basic attributes of self-definition. Nevertheless, private citizens who turn up in sexual memoirs find themselves in a catch-22: if they sue, they call attention to the book and, by making it additionally newsworthy, decrease their chances of legal recovery.

Because of changing sexual mores, even highly personal disclosures that might have been actionable a few decades ago are likely to be considered inherently newsworthy today. In 1942, for example, Marjorie Kinnan Rawlings, the author of the best-selling children's classic *The Yearling,* published an autobiography called *Cross Creek.* It contained some tame passages describing her horseback ride into the backcountry of Alachua County, Florida, with the local census taker, Zelma, whom Rawlings described as "an ageless spinster resembling an angry and efficient canary."[31] Rawlings depicted "my profane friend Zelma" as a lovably hardscrabble character with her own "special brand of profanity,"[32] and she quoted the census taker cursing boisterously at the backcountry inhabitants as she sympathized with their rough but charming lives. Zelma, whose last name Rawlings hadn't disclosed, was less than delighted by the portrait, and sued on the grounds that her privacy had been violated. The Supreme Court of Florida agreed, holding that Rawlings's celebrity was irrelevant to the question of the newsworthiness of the revelations: "No legitimate or general public interest in the defendant alone can justify an invasion of the right of privacy of another," the court held, "who has in no sense of the word consented to that invasion or waived her rights."[33]

But three decades later, a similar suit against a memoir writer ended very differently. In the 1970s, Will D. Campbell, a white civil rights leader, published an autobiography called *Brother to a Dragonfly*. In the course of describing his close relationship with his brother Joseph, he included details about the brother's home life and marriage, including passages suggesting that his brother's wife was unchaste and a liar, and that Joseph, a pharmacist, had beaten her while on drugs. The wife sued for invasion of privacy, but a federal court rejected her suit. "The story of a pharmacist's addiction to drugs from his pharmacy shelves might alone justify an intrusion into his wife's anonymity," the court held, and the fact that Will Campbell was "a person in whom the public has a rightful interest" made the account of his close relationship with his older brother inherently newsworthy.[34] More recently, in 1993, a woman appeared on a TV talk show and revealed that her daughter had been raped by the woman's husband. A Texas court held that the adult daughter couldn't recover for invasion of privacy because the mother had a First Amendment right to publish her personal account of her family tragedy, even though it involved revealing private facts at her daughter's expense.[35]

In suits over public disclosure of private facts, courts ask not only whether the information is newsworthy, but also whether its publication would be "highly offensive to a reasonable person."[36] And it's this part of the legal test that makes Brandeis's vision of the right to privacy especially hard to resurrect today. In an age that is beyond embarrassment, it's rarely clear what a "reasonable person" would find highly offensive. For every Duchess of York who objects to being captured, by telephoto lens, in a topless romp, there's a Jennifer Ringley, a twenty-one-year-old exhibitionist in Washington, D.C., who has a Web camera trained on her bedroom twenty-four hours a day. Who's the more reasonable person, Fergie or Jenny? "In everyday life, if you violate your friends' privacy you're not going to have many friends," says Robert Post, a professor of law at the University of California at Berkeley who has written provocative articles on privacy. "But in the social sphere, there's a sense in which we love to see the norms violated, and gossip is one way of defining community."

Post has written about the "civility norms" that Brandeis and Warren hoped to protect when they argued that invasion of privacy should be recognized as a legal wrong. The Boston lawyers sought to vindicate communal norms of reticence and ceremony as much as to redress the mental anguish suffered by individuals whose privacy had been violated.[37] Following Brandeis and Warren's suggestion, courts today agree that "community mores" should determine the line between information to which the public is entitled and "a morbid and sensational prying into private lives for its own sake," which may be an actionable invasion of privacy. But because community mores today are contested and unstable, the legal tests have proved far easier to apply in theory than in practice.

The legal test for invasions of privacy, unsurprisingly, looks very much like the legal test for obscenity, which focuses not only on whether the work, taken as a whole, lacks serious social value, but also on whether "the average person, applying contemporary community standards, would find that the work, taken as a whole, appeals to the prurient interest" and depicts sexual conduct in a "patently offensive way."[38] (To be obscene, the joke goes, a work has to turn you on and gross you out at the same time.) The trouble, as courts have found in cases involving the transmittal of salacious material over the Internet, is that community standards are increasingly hard to discern in an age in which self-exposure has gotten the better of reticence, and in which reading and writing are increasingly taking place in virtual communities whose values may be very different from those of the geographic communities in which individual members happen to reside.

In order to be actionable today, the disclosure of embarrassing facts has to be so outrageous that it verges on voyeurism or obscenity, leaving no doubt that the exposure would shame and humiliate a person of ordinary sensibilities in any community. In 1998, for example, Pamela Anderson Lee and her former lover, Bret Michaels, of the rock group Poison, sued to enjoin an Internet site from broadcasting an explicit videotape of Lee and Michaels having sexual intercourse. (The tape became public after Michaels unwisely presented it as a gift to a colleague, who proceeded to sell it to a pornographer.) The suit might be seen as the definition of chutzpah, since Lee had previously released videotapes of herself performing similar acts with her former husband, Tommy Lee. But a California court aptly noted that even sex symbols have privacy: "Public exposure of one sexual encounter [does not] forever remove a person's privacy interest in all subsequent and previous sexual encounters," the Court announced with a Brandeisian flourish. "Sexual relations are among the most private of private affairs, and . . . a video recording of two individuals engaged in such relations represents the deepest possible intrusion into such affairs."[39] In addition to violating Michaels and Lee's right to privacy, the judge held, broadcasting the video would also violate their right to publicity—namely, the right to sell their sex tape to the highest bidder of their choice. These days, the reticence team should be grateful for small victories.

The legal protections for privacy at home that the courts

have allowed to atrophy over the past few decades presuppose that society can distinguish "reasonable" from "unreasonable" searches, and can agree about what kind of sexual exposure would offend a "reasonable" person. Now that jurors, citizens, editors, and journalists are no longer confident about their ability to make those judgments, perhaps it's too much to ask that judges should make them for us. There is, in other words, a little Kenneth Starr in all of us, an inquisitorial imp eager to sift through diaries and listen to sex tapes, even as it purports to abhor them. But can we devour tabloid accounts of Princess Diana's cell phone intimacies and then deplore invasive questions about sex? "If we don't sanction people socially who violate privacy, we can't expect the law to do much of that work for us," Post says. "The point is that we should actually care about privacy, and it's not clear that we do."

Chapter 2

Privacy at Work

In response to a subpoena from Kenneth Starr's grand jury, Monica Lewinsky's best friend, Catherine Allday Davis, turned over the laptop computer that she had used to send e-mails to Lewinsky. During an interview with the FBI, Davis mentioned that, before she received the subpoena, she had tried unsuccessfully to erase the e-mails because she didn't want them to fall into the hands of the press and felt the need to protect her privacy. (She also worried that she would get into trouble for sending personal e-mail from a computer that belonged to her husband's employer.) Faced with the possibility of being accused of trying to destroy relevant evidence in an investigation of which she was only faintly aware, Davis entered into an immunity agreement with Starr's office on her lawyer's advice. Starr then recovered the deleted e-mails from the hard drive of the subpoenaed computer and printed them out without regard to their relevance to the case. Ignoring the protests of Davis and her lawyer, Starr's deputies included in their report to Congress Davis's private correspondence, which contained an intimate account of her feelings about her young husband during their honeymoon in Tokyo. "I have never been so angry in my life," she told Andrew Morton. "It was so violating to have your personal thoughts published for the world to read."[1]

The resurrection of long-forgotten e-mails is becoming a staple of courtroom dramas as well. Eighty-five percent of the evidence in the Iran-contra hearings is said to have come from exhumed e-mail.[2] And in 1997, Judge Thomas Penfield Jackson, the same judge who had ordered Bob Packwood to produce his diaries, chose Lawrence Lessig of Harvard Law School to advise him in overseeing the antitrust dispute between the government and Microsoft. When Microsoft challenged Lessig's appointment as a "special master," Netscape officials turned over to the Justice Department an e-mail that Lessig had written to an acquaintance at Netscape in which he joked that he had "sold my soul" by downloading Microsoft's Internet Explorer. The Justice Department, in turn, gave Lessig's e-mail to Microsoft, which claimed he was biased and demanded his resignation.

In fact, Lessig's e-mail joke had been quoted out of context. As the full text of the e-mail makes clear, Lessig had downloaded Microsoft's Internet Explorer in order to enter a contest to win a PowerBook computer. After installing the Explorer, he discovered that his Netscape bookmarks had been erased. In a moment of frustration, he fired off the e-mail to the Netscape acquaintance, whom he had met at a cyberspace conference, describing what had happened and quoting a Jill Sobule song that had been playing on his car stereo: "sold my soul, and nothing happened." And although a court ultimately required Lessig to step down as special master for technical reasons having nothing to do with his misinterpreted e-mail, he discovered that strangers were left with the erroneous impression that the e-mail "proved" that he was biased and forced him to resign. The experience taught Lessig that, in a world in which most electronic footsteps are recorded and all records can be retrieved instantly, it's very easy for sentiments to be misinterpreted by people with short attention spans who are too busy to put them in a broader context.

"The thing I felt most about the Microsoft case was not the actual invasion (as I said, I didn't consider it an invasion)," Lessig e-mailed me after the ordeal. "What I hated most was that the issue was just not important enough for people to understand enough to understand the truth. It deserved 1 second of the nation's attention, but to understand it would have required at least a minute's consideration. But I didn't get, and didn't deserve, a minute's consideration. Thus, for most, the truth was lost." Lessig felt ill-treated, in short, not because he wasn't able to explain himself, but because, in a world of short attention spans, he was never given the chance. This chapter will examine the ways that new technologies of communication at work—from e-mail to Internet browsing—increase the risk that our private letters and jokes can be taken out of context and misunderstood. It will also examine changes in the legal protections for privacy at work that have increased the risk that all of us may be judged unfairly by strangers and employers who don't have the time to understand the difference between our public and private speech.

Privacy can be conceived along two dimensions. There is the part of life that can be monitored, or directly observed, and there is the part of life that can be searched, because it leaves permanent records.[3] In the eighteenth century, technologies of monitoring and searching were primitive: in small villages, neighbors could observe each other in public, and social norms were strictly enforced through gossip, but, aside from private diaries and letters, there were few permanent records of the activities of daily life. This created a meaningful sphere of privacy in the home. Privacy, in an age of primitive technology, was largely a function of inefficiencies in technology of monitoring and searching.

In an electronic age, those inefficiencies have dramatically decreased, as it becomes possible to monitor someone's activities at the computer in his or her bedroom without breaking down the bedroom door. Most of us, if we are lucky, will never be caught up in lawsuits that lead to subpoenas for our computers or e-mail. But many of us have our e-mail and Internet browsing monitored by our employers, whether we know it or not. A survey of nearly a thousand large companies conducted by the American Management Association in 1999 found that 45 percent monitored the e-mail, computer files, and phone calls of their workers.[4] Nearly two-thirds engaged in some form of electronic surveillance in the workplace,[5] many without warning. Other estimates of the percentage of employers who monitor their workers' computer files, Internet use, and e-mail range from less than 10 percent to more than one-third.[6] Some companies use computer software that can monitor and record every keystroke on the computer with granular precision, such as a program called Spector, which takes digital snapshots of whatever appears on the screen. Other employers use an even more sophisticated program, called Assentor, that screens every incoming and outgoing e-mail for evidence of racism or sexism or body parts. After assigning each e-mail an offensiveness score, the program forwards messages with high scores to a supervisor for review.[7] At the law school where I teach, an e-mail censorship program was mistakenly installed on computers in the library. Students who typed the word "drugs" on their university e-mail accounts before the program was deactivated were automatically logged off the network with a stern warning.

At the end of the nineteenth century, Louis Brandeis wrote confidently that not only the contents of private letters but even a general description of their contents should be immune from publication. How, then, do we find ourselves in a world in which employers and universities can search their employees' e-mail and monitor their reading habits on the Internet with impunity? Part of the answer has to do with the Supreme Court's response to the growth of new technologies of monitoring and surveillance, which has proved to be distressingly passive at every turn. The Court's first encounter with electronic searches was a case decided in 1928 called *Olmstead v. United States,* which arose when the federal government began to tap phones in an effort to enforce Prohibition. Olmstead was the general manager of a fifty-man enterprise that produced $2 million a year by importing and selling illegal liquor transported by two ships between Seattle and British Columbia. Federal agents put wiretaps on the phone lines near Olmstead's home and office, and at his subsequent trial testified about the conversations they overheard. Olmstead claimed that the wiretaps violated the Fourth Amendment, which protects the right of the people to be secure in their "persons, houses, papers, and effects," but in a literal-minded opinion, Chief Justice William Howard Taft disagreed. The Fourth Amendment, he said, was originally understood to forbid only searches or seizures accompanied by physical trespass. The agents had not trespassed on Olmstead's property when they placed wiretaps on the phone lines in the streets near his house. Moreover, the conversations that the agents overheard were not material things, like papers or other personal property, and therefore were not protected by the Fourth Amendment.

In a visionary dissenting opinion, Louis Brandeis, who had become an associate justice in 1916, grappled with the issue of adapting late eighteenth-century values for a twentieth-century world. When the Constitution was adopted, Brandeis wrote, breaking and entering into the home was the only way for the government to invade a citizen's private thoughts. But in the 1920s subtler ways of invading privacy, such as wiretapping, had become available to the government. As private life had begun to be conducted over the wires, sealed letters were no different from private telephone messages, and the Supreme Court had held in 1877 that sealed letters and packages couldn't be opened by postal inspectors without a judicial warrant. Moreover, Brandeis noted, wiretaps invade the privacy of people on both ends of the line, exposing far more intimate activity than the general warrants of the eighteenth century. To protect the same amount of privacy that the framers of the Fourth and Fifth Amendments intended to protect, Brandeis concluded, it was necessary to extend those amendments to prohibit warrantless searches and seizures of conversations over the wires, even if the invasions occurred without physical invasions.[8]

Looking forward to the age of cyberspace, Brandeis predicted that technologies of surveillance were likely to progress far beyond wiretapping. "Ways may some day be developed by which the Government, without removing papers from secret drawers, can reproduce them in court, and by which it will be enabled to expose to a jury the most intimate occurrences of the home," Brandeis observed. In the Wilkes trial, "a far slighter intrusion seemed 'subversive of all the comforts of society.'" How, then, could the Constitution be interpreted to afford no protection against government monitoring of secret papers and letters sent over telephone wires, rather than forcibly extracted from a desk drawer?

A few decades later, Brandeis had his answer. In 1967, the Supreme Court appeared to accept Brandeis's argument that technologically enhanced eavesdropping could qualify as an unreasonable search, but it did so in a way that inadvertently undermined Brandeis's central insight. In *Katz v. United States*, government agents had attached a listening device to a public telephone booth and recorded a gambling suspect's end of the conversation without his knowledge. Overruling the *Olmstead* decision, which had held that there could be no search or seizure without a "physical intrusion," the Court announced that the "Fourth Amendment protects people, not places." Because Mr. Katz, the suspected gambler, had taken steps to preserve his privacy by closing the door of the phone booth behind him, the Court held that he reasonably expected his conversations wouldn't be monitored without a judicial warrant. In an influential concurring opinion, Justice John Marshall Harlan proposed the following test for determining what kind of surveillance activity should trigger the protections of the Fourth Amendment: a person must have an actual or subjective expectation of privacy, Harlan suggested, and the expectation must be one that society is prepared to accept as reasonable.[9]

Harlan's test was applauded as a victory for privacy, but it soon became clear that it was entirely circular. People's subjective expectations of privacy tend to reflect the amount of privacy they subjectively experience; and as advances in the technology of monitoring and searching have made ever more intrusive surveillance possible, expectations of privacy have naturally diminished, with a corresponding reduction in constitutional protections. In a series of related rulings, the Court held that if you share information with someone else, you relinquish all "reasonable expectation of privacy" that the information will remain confidential. In the 1971 case that made it possible for Kenneth Starr to wire Linda Tripp, four justices said that a government informer carrying a radio transmitter could secretly broadcast his conversation with a suspected drug dealer to an agent waiting in a nearby room, because all of us, when we confide in our friends, assume the risk that our friends may betray us. And, in the cases that laid the groundwork for Starr's subpoenas of Monica Lewinsky's bookstore receipts, the Burger Court decided, in the 1970s, that we have no expectation of privacy in information such as bank records and telephone logs that we voluntarily turn over to a third party. The Court insisted, again, that when we share information with other people, all of us assume the risk that those people may disclose the information to the government.

The justices' sweeping generalizations in these cases bear little relation to the kind of privacy that citizens expect in the real world. When Americans reveal information to a bank, they don't, in fact, believe that the bank will turn their deposit information over to the government, as Congress recognized when it passed a law limiting the effect of the bank records decision by prohibiting banks from turning over records to federal agencies. In 1993, moreover, two law professors, Christopher Slobogin and Joseph E. Schumacher, decided to test the Supreme Court's generalizations by interviewing nearly two hundred citizens about their expectations of privacy.[10] Asked to rank fifty hypothetical searches conducted by the government without consent, in ascending order of how much the searches seemed to intrude on citizens' expectations of privacy, the survey participants listed "perusal of bank records" in thirty-eighth place—only ten places below the four most intrusive searches of all: number 47, the search of a bedroom; number 48, reading a personal diary; number 49, monitoring the phone for thirty days; and number 50, a body cavity search at the border.[11] According to the survey, the Supreme Court was also wrong to suggest that the government can recruit undercover agents to engage in secret taping without violating "justifiable expectations of privacy" because all of us assume that our confidants may betray us. The use of a chauffeur and a secretary as undercover agents ranked thirty-first and thirty-fourth on the intrusiveness scale, surpassing even "arrest, handcuffing, and detention for 48 hours."[12]

Finally, the Court has held that government employers are free to search the offices of their employees for work-related misconduct, without clear suspicion of wrongdoing, on the theory that people expect less privacy in workplaces that they share with other employees.[13] But Slobogin and Schumacher found that "going through drawers at [the] office" ranked twenty-seventh on their intrusiveness scale, considerably above a frisk on the street, which usually requires that the police have reasonable suspicion that someone is armed.[14] The Court has also held that certain government agencies can conduct random drug and alcohol testing of its employees, on the theory that railway workers and customs agents have diminished expectations of privacy.[15] But Slobogin and Schumacher found that urinalysis and blood extraction with a needle were considered quite intrusive, in thirty-ninth and forty-sixth place respectively.[16]

It's not surprising that Supreme Court justices, who are secluded in a marble palace and have spent most of their careers in the cosseted solitude of lower courts and universities, aren't terribly good at predicting how much privacy ordinary Americans expect in the workplace. The Court's reasoning—that a person who confides in someone else must run the risk that his secret will be betrayed—is simplistic at best: in the bank records case, the bank managers hadn't *chosen* to betray the confidences of their depositors. In fact, the government had ordered the bank to keep records of deposits and then *forced* the bank to disclose those records to a federal grand jury.[17] If the Court meant what it said—"a person has no legitimate expectations of privacy in information he voluntarily turns over to third parties"[18]—then it would have to reconsider its holding in the wiretapping case, where it said that a person does have a legitimate expectation of privacy in information shared with a friend on the telephone. But the real problem with the Supreme Court's test for invasions of privacy is not empirical but conceptual. In many cases, people have an objectively valid expectation of privacy that the Court, by judicial fiat, has deemed unjustifiable. We need more independent mechanisms for protecting privacy—such as grand juries or other popularly accountable bodies—which can balance the claims of the police against the privacy of individuals and decide whether a search is reasonable before information is disclosed.

Congress might well decide to guarantee more privacy for bank records than the Constitution requires. But it makes no sense for the Court to say that when I reveal private information in one context, I relinquish the right to conceal the same information in other contexts. As Oliver Sipple argued in vain, privacy should allow us to reveal parts of ourselves in one context that we conceal in another context. And a core principle of informational privacy, widely recognized in Europe, holds that personal information gathered for one purpose shouldn't be disclosed, made available, or otherwise used for another purpose without the consent of the individual concerned.

In an important dissent in the bank records case, Justice Thurgood Marshall noted that constitutional protections for privacy shouldn't turn on subjective expectations, which necessarily diminish as technologies of surveillance permit the state to invade privacy in more efficient, but less detectable, ways: "Whether privacy expectations are legitimate," Marshall wrote, "depends not on the risks an individual can be presumed to accept when imparting information to third parties, but on the risks he should be forced to assume in a free and open society."[19] A vision of privacy that took seriously the text of the Fourth Amendment might emphasize that there is an irreducible core of constitutional protection against unreasonable searches and seizures of persons, houses, electronic papers, and effects that is necessary for freedom, regardless of how much or how little privacy people subjectively expect in these areas in the light of changing technologies of surveillance.[20]

Journalists, authors, and lawyers are the only people who enjoy a measure of protection for private papers in the workplace today, because in cases involving searches of newspaper and law offices, Congress has rebelled at the Supreme Court's suggestion that private papers that are stored at work can be searched and seized with impunity. In 1971, for example, *The Stanford Daily* published pictures of a violent clash between student protesters and local police officers during a demonstration at the Stanford University Hospital. After receiving a warrant, the police officers proceeded to search the student newspaper's office for all negatives and photographs taken at the demonstration, combing through file cabinets, desks, darkrooms, and trash cans. *The Stanford Daily* then sued the police department, arguing that the police shouldn't be able to search for evidence in the possession of someone who wasn't suspected of a crime, unless a subpoena would be futile. The Supreme Court, in 1978, breezily dismissed the newspaper's claim, leaving no doubt that the nineteenth-century doctrine that prohibited searches for "mere evidence" had been repudiated. But the idea that the police could invade newspapers in search of photographs and reporters' notes was so offensive to defenders of free speech that Congress responded by passing the Privacy Protection Act of 1980, which says that "[n]otwithstanding any other law, it shall be unlawful for a government officer or employee, in connection with the investigation . . . of a criminal offense, to search for or seize any work product materials possessed by a person reasonably believed to have a purpose to disseminate to the public a newspaper, book, broadcast, or other similar form of public communication."[21]

The Privacy Protection Act has created an enclave of privacy for the small group of people who can plausibly present themselves as journalists or publishers. In a small victory in cyberspace, for example, Steven Jackson Games, an electronic publisher and purveyor of role-playing games in Texas, successfully challenged the U.S. Secret Service's search of his office and seizure of the computer responsible for running an electronic bulletin board system. The bulletin board posted information about the exotic role-playing games that the company sold—many, appropriately enough, involved cyberpunk fantasies about gothic conflicts between humans, the state, and technology—and offered a free e-mail service for its subscribers. The case began when Bell South, the local telephone company, told the Secret Service that a program identifying the employees responsible for operating its 911 emergency system had been stolen by a group of computer hackers, who called themselves the Legion of Doom. (Far from being top secret, the program turned out to have been available to the public for twenty dollars.) One of the hackers who received the document was a man called Lloyd Blakenship, and the Secret Service got a warrant to search the office of his employer, Steven Jackson Games, even though it had no reason to suspect that the company itself had violated the law. In 1990, the Secret Service raided the Steven Jackson Games office and carried away three computers, three hundred computer disks, and the unpublished draft of a book called "GURPS Cyberpunk." As a result of the raid, the bulletin board nearly went out of business, and the stored e-mails of several customers were read and accidentally deleted. After the raid, Steven Jackson Games sued the Secret Service, and a Texas district court agreed that because the company was a publisher as well as a game manufacturer, the raid had violated the Privacy Protection Act. The court awarded more than fifty thousand dollars in expenses and damages.[22]

Although employees of publishing companies now have a modicum of protection against broad searches of their private papers, those in other professions are out of luck. The ironic consequence of Congress's solicitude for the rights of journalists is that someone who stores private papers or e-mails at work with no intention of publishing them has *less* privacy protection than someone who intends eventually to publish his thoughts. The leading Supreme Court decision about workplace privacy was handed down in 1985, and it gave public employers broad discretion to search the private papers of their employees on the flimsiest of suspicions.[23]

The case began, again, with a controversy about a computer. In 1981, Dr. Magno Ortega, a psychiatrist who was chief of professional education at Napa State Hospital, bought a new computer to help train his residents. He paid half the cost of the computer and the residents donated money to cover the rest. The executive director of the hospital, Dr. Dennis O'Connor, then began to ask questions about whether the computer had been properly donated to the hospital, and whether the residents had been coerced into paying for it. O'Connor started to investigate Ortega, and suddenly other vague allegations of misconduct surfaced.[24] One woman, a current resident, said that Ortega made her unhappy at work and one Saturday morning showed up at her house. Another woman, a former resident, said that in the 1970s, about a decade earlier, Dr. Ortega had "stroked her hair" and said that she was "very special and had to leave her husband."[25] Neither woman used the term "sexual harassment" or filed formal harassment charges, although in her subsequent Supreme Court opinion, Justice Sandra Day O'Connor seemed to put great weight on the hospital's claim that it was investigating charges relevant to sexual harassment when it put Ortega on administrative leave and proceeded to search his office without a warrant.[26]

In a highly intrusive search that evoked the indignities visited on John Wilkes, the investigators combed through Ortega's locked desk drawers and his private file cabinets. In his desk drawers the investigators found and read personal letters from his friends, his ex-wife, and his daughter, as well as sexually explicit letters from several female friends over many years. Without distinguishing between private and public property, the investigators seized the letters, as well as photographs of family members, appointment books, and the manuscript of an unpublished book.[27] The business manager of the hospital, Dr. Richard Friday, returned to conduct an exploration of his own. He removed from Dr. Ortega's desk drawer a valentine, a suggestive photograph, and an inscribed book of poetry that Ortega had received about ten years earlier from a former resident, Dr. Joyce Sutton. These objects were later produced to embarrass Dr. Sutton when she testified on Ortega's behalf at a state personnel board hearing convened to discredit him.[28]

Soon after this invasive search, Dr. Ortega filed a complaint against Dr. O'Connor and Dr. Friday, arguing that they had conducted an unreasonable search and seizure of his office and private papers, in an effort to find embarrassing material that might be used against him. The case made its way up to the Supreme Court, where it produced a plurality opinion. Writing for the court, Justice O'Connor held that the amount of privacy that employees can legitimately expect in different parts of the workplace should vary in accordance with how much privacy they subjectively experience. "Public employees' expectations of privacy in their offices, desks, and file cabinets," she held, "may be reduced by virtue of actual office practices and procedures, or by legitimate regulation."[29] O'Connor assumed that Ortega had a reasonable expectation of privacy in his desk drawers and file cabinets, because he didn't share those areas with any other employees. But she held that searches of even the most private areas in Ortega's office didn't require a judicial warrant or probable cause to believe that he was guilty of some specific wrongdoing. The search of Ortega's desk drawers should be upheld, O'Connor said, as long as both the initial justification for the search and its subsequent scope were reasonable. Since Ortega had taken his computer home, she said, the state could legitimately search throughout his office to see whether he had removed any other state property. "The removal of the computer—together with the allegations of mismanagement of the residency program and sexual harassment—may have made the search reasonable at its inception." She then sent the case back to the lower court to determine whether or not the scope of the search was, in fact, reasonably related to its ostensible purposes.[30]

In 1997, sixteen years after the initial search, a jury found that the hospital had conducted an unconstitutional fishing expedition aimed at turning up incriminating evidence against Dr. Ortega rather than a reasonable search to inventory government property. It awarded him $376,000 in compensatory damages and $60,000 in punitive damages. Even more significantly, the trial judge ruled that evidence of Dr. Ortega's private papers— including the photograph, the valentine, and the book of love poetry from Dr. Sutton—couldn't be presented at the trial, because one of the "sexual harassment" allegations was too vague and the other was too old to have justified such an invasive search.

In agreeing that all evidence of sexual harassment should be excluded, the appellate court showed a sensitivity to the special status of private papers that the Supreme Court had not. The Ninth Circuit wrote:

> Given the privacy interests involved, it has long been apparent that stale and unsubstantiated allegations do not entitle supervisors to rummage through employees' desks and file cabinets without a reasonable belief that specific evidence of misconduct will be found, let alone scrutinize and seize whatever personal or romantic letters or mementos they may happen to locate. . . . A charge that a person has engaged in what may be offensive sexual conduct on one or more occasions does not make his or her entire personal and family existence

fair game for indiscriminate and unrestrained governmental intrusion and examination.[31]

Dr. Ortega may have been belatedly vindicated, but the Supreme Court opinion in his case has created an incentive for employers to search the most private areas of the workplace as regularly as possible, in order to decrease their employees' expectation of privacy. And this leaves personal e-mail sent over a company server especially vulnerable to exposure. Personal e-mail poses something of a constitutional puzzle: on the one hand, it sometimes originates from work, which is not a place that the Fourth Amendment reserves for special constitutional protection; on the other, it seems analytically similar to a private letter, which is one of the "papers" at the historical core of the Fourth Amendment. It's true that e-mail, when not encrypted, can be intercepted relatively easily: rather than being sent directly from computer to computer, the data travels in packets through several intermediaries, where it can be read before it arrives at its destination. Moreover, e-mail can be retrieved from several destinations after it arrives, including the hard drive of the sender or the recipient, or the employer's network computer, where it is often stored on backup tapes even after being deleted.

But the fact that e-mail can be intercepted physically doesn't mean that it should be treated, for legal purposes, as if it were a postcard. In colonial America, letters from Europe were left at local taverns by ship captains, open for public inspection until they were claimed. And at the end of the eighteenth century, around the time of the framing of the American Constitution, the mail was so insecure that Postmaster General Benjamin Franklin and, later, Thomas Jefferson thought that their own mail was being opened. (Indeed, Jefferson invented an early encryption machine to address this problem.) To alleviate similar concerns, Congress passed the Postal Act of 1825, which prohibited prying into other people's mail.[32] And in 1877, as I mentioned, the Supreme Court held that government needed a search warrant to open first-class mail, regardless of whether it was sent from the office or from home.

In cases involving e-mail sent from home over commercial Internet providers, courts have begun to recognize the analogy between sealed letters and personal e-mail. "In a sense, e-mail is like a letter," a military court noted in a recent case involving the investigation of an Air Force colonel who had been tried in 1993 for sending "indecent" private messages. In the messages, which the colonel had sent from home, while off duty, on a personal AOL account, he confessed his confusion about his sexual orientation to a friend. All e-mails sent over AOL are stored at America Online's headquarters in Virginia, but this didn't change the court's conclusion that each e-mail "lies sealed in the computer until the recipient opens it" and therefore "the transmitter of an e-mail message enjoys a reasonable expectation that police officials will not intercept the transmission without probable cause and a search warrant."[33]

Why is it, then, that in cases involving e-mail sent from work, courts are increasingly holding that employees have very little expectation of privacy? Mostly because of the tautological argument that Justice O'Connor endorsed in the Ortega case: as long as network administrators have the technical ability to read their employees' e-mail, employees should have no reasonable expectation that their e-mails aren't being read. In 1996, for example, police officers from Reno, Nevada, objected that their Fourth Amendment rights were violated when e-mail messages they had sent over the department's internal message system were retrieved from a central computer. A court rejected their claim, quoting a commentator who noted that "an employee's privacy interest in E-mail messages" would likely "fail the 'expectation of privacy' test since most users probably realize that a system administrator could have access to their E-mail."[34]

Courts have reached similar conclusions in invasion of privacy suits filed against private employers who fired employees after reading their e-mail. (Private employers are not bound by the restrictions of the Fourth Amendment, but the legal tests for unreasonable searches and seizures in public workplaces and invasions of privacy that take the form of "intrusion on seclusion" in private workplaces are similar.) The plaintiffs in one of the cases, Bonita Bourke and Rhonda Hall, had been hired to set up an e-mail system for a Nissan dealership. In the course of demonstrating how the system worked, a coworker randomly picked a personal e-mail that Bourke had sent, which turned out to be sexually explicit. The Nissan management proceeded to review all of the messages sent by Bourke and Hall and reprimanded both of them after finding several personal messages that contained sexual banter. When Bourke and Hall protested, Hall was fired and Bourke resigned. Bourke then sued, arguing that Nissan had invaded her privacy by reading her e-mail, but the California Court of Appeals rejected her claim, concluding that Bourke didn't have a reasonable expectation of privacy in her e-mail because she knew months earlier that Nissan was monitoring e-mail messages. In addition, Bourke had signed a statement that read: "It is company policy that employees and contractors restrict their use of company-owned computer hardware and software to company business."[35]

Cases like this suggest that merely by adopting a written policy that warns employees that their e-mail may be monitored and restricted, employers will lower expectations of privacy in a way that gives them even broader discretion to monitor and restrict their employees' e-mail. "An employer should develop a policy that effectively lowers the expectation of privacy in advance," as one commentator crisply puts it. "This will greatly improve an employer's chances of tipping the privacy balance in its favor in future litigation challenging the surveillance or monitoring."[36] In general, intrusion on seclusion suits against private employers rarely succeed, except in the most extreme cases, involving monitoring of locker rooms and bathrooms, because employees are hard-pressed to prove that the workplace is a sufficiently private area to create reasonable expectations of privacy, or that the monitoring would be "highly offensive to a reasonable person."[37]

If employers were permitted to monitor their employees' e-mail only after clearly warning the employees in advance to expect monitoring, then the surveillance might be tolerated as an intrusive but freely accepted condition of employment. It's arguable, after all, that if employers were prohibited from conducting searches to investigate violations of workplace rules, which the Supreme Court has suggested is nearly always reasonable,[38] then they might hesitate to make private spaces available to their employees in the first place—moving file cabinets into hallways, for example, or forbidding locks on drawers, or requiring that all correspondence be stored in communal, rather than private, computer files.[39] Indeed, the courts might adopt the following default rule: if the employer doesn't expressly warn employees of the specific invasions of privacy that the company practices, and gain their consent, then the material in question can't be invaded.

Unfortunately, judges today have adopted something like

the opposite rule: even when employers promise to respect the privacy of e-mail, the employers have a right to break their promises without warning. In a recent case from Pennsylvania, for example, Pillsbury repeatedly promised its employees that all e-mail would remain confidential and that no employee would be fired on the basis of intercepted e-mail. Michael Smyth, a Pillsbury employee, received an e-mail from his supervisor over the company's computer network, and he read the e-mail at home. Relying on the company's promise about the privacy of e-mail, he shot back an intemperate reply to the supervisor, allegedly saying at one point that he felt like "kill[ing] the back-stabbing bastards" on the sales force, and referring to an up-coming holiday party as the "Jim Jones Koolaid affair."

Despite its promises, the company proceeded to retrieve from its computers all the private e-mails that Smyth had sent and received during the month of October. It then fired him for trans-mitting "inappropriate and unprofessional comments." Smyth sued, arguing that the company had invaded his right to privacy by firing him. But the court blithely dismissed his claim. Once Smyth voluntarily "communicated the alleged unprofessional comments to a second person (his supervisor) over an e-mail system which was apparently utilized by the entire company, any reasonable expectation of privacy was lost," the court an-nounced. Moreover, even assuming that an employee has a rea-sonable expectation of privacy when he uses a company e-mail system, the court added, a reasonable person wouldn't consider the interception of e-mail to be "a substantial and highly offen-sive invasion of . . . privacy." Pillsbury's "interest in preventing inappropriate and unprofessional comments or even illegal ac-tivity over its e-mail system outweighs any privacy interest the employee may have in those comments."[40]

This can't be right. The privacy interests that are invaded by reading private e-mail are very high indeed. The sociologist Georg Simmel has written about the ways in which written letters are peculiarly subject to misinterpretation. Because letters lack the contextual accompaniments—"sound of voice, tone, gesture, facial expression"—that, in spoken conversation, are a source of both obfuscation and clarification, Simmel argues, letters can be misinterpreted more easily than speech, and "what in human utterances is clear and distinct, is more clear and distinct in the letter than in speech, and what is essentially ambiguous, is more ambiguous."[41] With e-mail, which captures a range of highly subjective and informal private emotions, the possibilities for misinterpretation are even more acute. E-mail combines the intimacy of the telephone with the retrievability of a letter. (Indeed, e-mails are far more retrievable than letters because they are collected in one place and are hard to delete.) Messages sent by e-mail are often far more impetuous than face-to-face conversations, where "situational cues," such as body language and facial expressions, from the person with whom we are conversing temper what we say. Because e-mail messages are often dashed off quickly and sent immediately, without the opportunity for second thoughts that ordinary mail provides, they may, when wrenched out of context, provide an inaccurate window on someone's emotions at any particular moment. Many of us use what amounts to a private vocabulary in our e-mail with close friends, a vocabulary informed by shared experiences and background assumptions. But these unspoken assumptions are inaccessible to outsiders, and can easily lead to misunderstandings about whether a message is meant to be whimsical or serious.

Lawrence Lessig's joke is a jarring example of the dangers of misinterpreting e-mail jokes when they are wrenched out of context. And after complaining to his acquaintance at Netscape that Microsoft's Internet Explorer had eaten his bookmarks, Lessig concluded: "When are you in Cambridge? I've moved permanently now, and I'd love to show you the center, and talk and whatever else we can think of doing." A newspaper in the Northwest later pointed to Lessig's suggestion that he and the Netscape lawyer should do "whatever else we can think of doing" and speculated absurdly that they were homosexual lovers. Like Smyth's comment about killing someone, Lessig's e-mail was subject to several possible interpretations, but most of them were silly when evaluated in context. The problem with e-mail is that the text is never accompanied by its original context. Even among friends, e-mail is often misunderstood: many of us have had the experience of writing a serious message that is perceived as ironic, or an ironic message that is taken too seriously, because e-mail shares the informality of a conversation, but like a letter, lacks the contextual accompaniments that provide clues to meaning in face-to-face encounters.

On any company network, a user's e-mail folders of sent and received messages are likely to contain a range of public and private expression, from the official statements of company policy that are traditionally found in letters to the private jokes and flirtation that used to take place around the watercooler. Even if a company reserves the right to monitor e-mail, few people really expect that their frivolous jokes will be exposed to the world, just as only paranoid people expect that their phones at work are being tapped. Of course, a world in which e-mail is hard to delete and easy to retrieve is partly a consequence of current technology, and technology can change. A San Francisco company called Disappearing Inc. recently invented a form of self-destroying e-mail that uses encryption technology to make messages nearly impossible to read soon after they are received. When I send you a message, Disappearing Inc. scrambles the e-mail with an encrypted key, and then gives you the same key to unscramble it. I can specify how long I want the key to exist, and after the key is destroyed, the message can't be read without a code-breaking effort that would be prohibitively expensive except in the most serious criminal investigations.[42]

But should ordinary citizens be forced to resort to esoteric encryption technology like Disappearing Inc. every time they send an e-mail? This would represent a bovine surrender to technological determinism, putting us in the same category as citizens in the Soviet Union. A court that took informational privacy seriously could try to think creatively about how to allow companies to search for evidence of workplace misconduct without exposing all their workers' private correspondence to public view, even if the correspondence happens to be stored rather than destroyed. Remember that in the eighteenth century warrants couldn't be issued to search for "mere evidence" of a crime, only for contraband or the fruits and instrumentalities of a crime. Perhaps judges should resurrect a version of the "mere evidence" rule for cyberspace, allowing only the illegal activity revealed by the search—but not the e-mail messages themselves—to be introduced in court. Or perhaps privacy masters, of the kind I discussed in the last chapter, could separate e-mail used as the instrumentality of a crime, which could be admitted in court, from e-mail that might be considered mere evidence of a crime, which could be excluded. (In the case of serious crimes, such as felonies, the master might balance the suspect's interest in privacy against the state's need for the information.) Striking this balance would necessarily involve subjective judgments that would make certain kinds of workplace misconduct harder to expose, but eighteenth-century courts were willing to bear the costs of underenforcing laws that prohibited less serious offenses in the interests of protecting personal privacy.

This, however, is not a world where companies are free to respect the boundary between public and private expression. When the judge in the Pillsbury case emphasized an employer's "interest in preventing inappropriate and unprofessional comments or even illegal activity over its e-mail system," he was referring to the fact that inappropriate and unprofessional e-mails can be considered as evidence of illegal activity, or as illegal activity in themselves. In 1995, for example, a woman named Karen Strauss sued Microsoft, arguing that she had been the victim of sex discrimination when she was passed over for a promotion to be technical editor of the company's software journal. The Equal Employment Opportunity Commission found that the evidence failed to support Strauss's claim, since the successful candidate was more qualified than she was. Nevertheless, at the trial that followed, a federal district court in New York permitted Strauss to introduce offensive e-mail messages that had circulated around the office, including a message that Strauss had received containing a satire called "Alice in UNIX Land," an e-mail advertisement that her boss had forwarded to the entire office containing a product announcement for replacement "Mouse Balls," and two e-mail messages that the boss had forwarded to male staff members only, including a news report about Finland's proposal to institute a sex holiday and a parody entitled "A Girl's Guide to Condoms."[43] Dismissing Microsoft's argument that these were childish attempts at humor, taken out of context, the court held that the e-mail messages "could lead a reasonable jury to conclude that Microsoft's proffered reason for failing to promote Strauss is not the true reason for its actions."[44] There is no reason, of course, that a prudent employer shouldn't be able to prohibit its employees from circulating sophomoric and offensive jokes to all the men and none of the women in the office. But the possibility that isolated jokes might cast light on the reasons that Strauss wasn't promoted was surely outweighed by the costs of wrenching the jokes out of context and making them the subject of legal scrutiny.

This brings us to the heart of the debate about workplace privacy at the beginning of the twenty-first century: the creation of a liability regime where monitoring of employees' speech and behavior has become a matter of corporate self-interest. There are many reasons that public and private employers monitor their employees—ranging from concerns about corporate and international espionage to concerns about productivity. But a common justification offered by courts and management lawyers to justify employer monitoring is a fear of liability for sexual harassment.[45] "If sexual and sexist communications were allowed to flourish in the workplace, unchecked and uncensored, an employer may be liable for sexual harassment," warns an article called "Preventing Internet-Based Sexual Harassment in the Workplace." "The same software sold to parents to block children's access to sex-related Web sites can be used by employers to control its employees' activities."[46]

Some studies have indicated that over 20 percent of e-mail users have received sexually explicit e-mail in the workplace. Several employers, including Eastman Kodak and Hallmark Cards, have identified sexual harassment as the most serious form of e-mail abuse.[47] An FBI survey conducted in 1996 found that 31 percent of the 563 businesses responding had suffered financial losses from sexually harassing e-mail and Internet use.[48] By itself, of course, a single e-mail or a practical joke from a boorish supervisor or coworker wouldn't be sufficiently "severe or pervasive" to be illegal under the Supreme Court's definition of harassment. But a "hostile environment" can be created by nothing more egregious than a pattern of comments or jokes from different employees, and courts have imposed liability not only for sexually hostile jokes, but for jokes that simply have sexual themes.[49] As a result, some management lawyers advise companies to monitor e-mail and Internet use strictly. A sample e-mail policy developed by one employment lawyer, after warning employees that their e-mail may be subject to unannounced inspections, announces: "You should not use the XYZ e-mail for gossip, including personal information about yourself or others, for forwarding messages under circumstances likely to embarrass the sender, or for emotional responses to business correspondence or work situations."[50] The policy continues: "You may not use XYZ's e-mail system in any way that may be seen as insulting, disruptive, or offensive by other persons, or harmful to morale. Examples of forbidden transmissions include sexually-explicit messages, cartoons, or jokes; unwelcome propositions or love letters; ethnic or racial slurs; or any other message that can be construed to be harassment or disparagement of others based on their sex, race, sexual orientation, age, national origin, or religious or political beliefs."[51]

Most people are surprised to learn that sexual harassment law does not impose liability on sexual harassers. Instead, it puts the full weight of responsibility on their employers. And owing to the incentives created by this liability regime, prudent companies have little choice but to restrict a great deal of sexual expression that no jury would ultimately condemn.

To understand the sources of the legal pressures on companies to monitor e-mail and Internet browsing, we must briefly review the tortuous evolution of the law governing the liability of employers for the sexual harassment of their employees. In 1986 the Supreme Court held that employers can reduce the danger of being held liable for sexual harassment by establishing procedures "calculated to encourage victims of harassment to come forward."[52] Since then, almost 75 percent of companies with more than one hundred employees have adopted anti-sexual-harassment policies. Most of those policies look very much like the "Sample Antiharassment Policy" reproduced in Barbara Lindemann and David Kadue's *Sexual Harassment in Employment Law,* which first appeared in 1992.[53] After quoting the EEOC's definition of sexual harassment, the model policy goes on to say that "[s]exual harassment may include explicit sexual propositions, sexual innuendo, suggestive comments, sexually oriented 'kidding' or 'teasing,' 'practical jokes,' jokes about gender-specific traits, foul or obscene language or gestures, display of foul or obscene printed or visual material, and physical contact such as patting, pinching, or brushing against another's body."[54]

To protect themselves from the risk of expensive liability, companies face increasing pressure to launch formal investigations in response to allegations of relatively trivial offenses, monitoring and prohibiting far more speech than the law actually forbids. "Now is the time to refine your sexual harassment policy and training program so that it is in the spirit of your organization's values rather than written to the letter of the law," writes Rita Risser in a report in 1996 by Fair Measures Management Law Consulting Group.[55] "Your policy should go beyond what the law forbids. If you set your standards too low, one mistake by one supervisor could make you the next landmark case. Also, the EEOC accepts claims for conduct that clearly is not illegal. Since it's costly to respond to such claims, it's in an organization's best interest to minimize them."[56] Eager to avoid liability, employers are cracking down on e-mail jokes. The Principal Financial Group in Des Moines, Iowa, for example, recently fired a customer service representative who sent coworkers sexually explicit jokes, including "Top 10 Reasons Why Trick-or-Treating Is Better Than Sex." A state judge awarded him unemployment benefits, however, concluding that Principal hadn't warned him against personal use of the company's e-mail or proved that other employees felt harassed.[57]

In 1998 the Supreme Court increased the pressure on companies and universities to ban more expression than the law actually forbids. In two important opinions, it held that when a supervisor with immediate (or successively higher) authority over an employee creates a hostile work environment, the company is automatically liable. When there are no tangible job consequences, however, the company can avoid liability by showing that it adopted a sexual harassment policy "to prevent and correct promptly any sexually harassing behavior," and that the plaintiff failed to take advantage of the policy in filing a complaint.[58]

There is a curiously disembodied quality to the opinions by Justices Anthony M. Kennedy and David H. Souter. Both are based on the assumption that sexual harassment should be treated no differently than ordinary cases involving intentional misdeeds by employees in the workplace, such as willful indifference to the safety of a product. But Kennedy and Souter fail to acknowledge the obvious difference between sexual harassment and ordinary injuries in the workplace: it is often impossible for a supervisor or employer to know in advance whether or not offensive conduct is illegal, because of the amorphousness of the legal definitions of harassment itself. The Court is overconfident, therefore, when it suggests that harassment is "eminently foreseeable" and that prudent companies can take steps to prevent it, in the same way that they take steps to prevent other workplace injuries.

In a dissenting opinion in the employment harassment cases, Justice Clarence Thomas discussed the tensions between individual privacy and the Court's mandate that employers must bear the responsibility for the speech and conduct of their employees. "Sexual harassment is simply not something that employers can wholly prevent without taking extraordinary measures—constant video and audio surveillance, for example—that would revolutionize the workplace in a manner incompatible with a free society," he wrote.[59] Thomas would hold companies responsible for the offensive speech of their supervisors only when they are negligent—that is, when the companies actually knew about the harassment and failed to do anything to stop it, or when they should have known about it but failed to adopt policies designed to prevent it.

In a provocative defense of the Supreme Court's liability regime, Robert Post has suggested that the corporate workplace is a "managerial" sphere in which social relations are organized around principles of efficiency. For this reason, he argues, citizens should be willing to accept greater restrictions on their autonomy and their privacy in the workplace than they would tolerate in the public sphere, which is ideally a space of self-governance.

But surely this Taylorite vision of the modern workplace, rooted in principles of industrial organization from the 1920s, is hard to accept today. As e-mail, modems, and PCs break down the boundaries between work and home, there are progressively fewer private or public spaces for citizens to express themselves autonomously. The Internet has blurred the distinction between the home and the office, as Americans are spending more time at the office and are using company-owned computers and Internet servers to do their work from home. But as technology poses new challenges to geographic conceptions of privacy, courts have not been encouraged to think creatively about how to reconstruct zones of individual privacy and free expression. On the contrary, the Supreme Court has rigidly extended harassment law from private and public employers to more fluid spaces, such as municipal beaches and federally funded schools and universities, including residence halls and playgrounds, which exist to promote free expression rather than managerial control.

As harassment law has developed, it seems ill equipped to make the fine distinctions among workplaces that Post imagines. Surely public universities and high schools should be regulated by less restrictive behavior codes than corporate businesses. The purpose of a university is to promote freedom of thought, not to increase productivity, and holding schools to the same standards as corporations arguably poses a direct threat to the experience of teaching and learning. And yet fear of liability for sexual harassment has also created a powerful legal incentive for universities and even libraries to monitor the reading habits of their employees, students, and patrons in cyberspace, and to adopt filtering software to protect employees from the possibility that someone else may download material that offends them. In 1997, for example, the public library board in Loudoun County, Virginia, voted to adopt a "Policy on Internet Sexual Harassment," requiring that filtering software called X-Stop be installed on all library computers to limit the access of patrons to the World Wide Web. "Library pornography can create a sexually-hostile environment for patrons or staff," the policy explained. "Permitting pornographic displays may constitute unlawful sex discrimination in violation of Title VII of the Civil Rights Act."[60] Colonel Dick Black, the author of the policy and a library trustee, defended it in the following terms: "People can do certain things in the privacy of their own homes that they cannot do in the workplace. Now this is not limited strictly to libraries. But the courts have said that whether it's a public state facility or whether it's a manufacturing plant, people cannot deprive women of their equal access to those facilities and their equal rights to employment through sexual harassment."[61]

In a rare burst of sanity, a district judge in Alexandria, Virginia, had no hesitation in striking down the Loudoun County Public Library's policy as a clear violation of the First Amendment rights of library patrons to receive information. Opponents claimed that because of the imprecision of the filtering software, which searches for forbidden words, like "sex," X-Stop blocked access not only to pornography but also to the Quaker Home Page, the Zero Population Growth Web site, and a site for the American Association of University Women in Maryland.[62] In striking down the policy, the district judge noted that the First Amendment forbade public libraries to restrict access to constitutionally protected books on the basis of their content. The only evidence the library offered to suggest that its policy was necessary to avoid sexual harassment was a single incident in another Virginia library, where a patron complained she had seen a boy viewing what she thought were pornographic pictures on the Internet. In response to the incident, the library installed privacy screens on its Internet terminals; a librarian testified that the screens "work great."[63] The district court noted that there were any number of other ways that Loudoun County libraries could shield employees from pornography without reducing adults to reading only what is fit for children—installing filtering software on some Internet terminals, for example, and allowing minors to use only those terminals, just as the library staff traditionally shooed away "prepubescent boys giggling over gynecological pictures in medical books."[64] Because filtering the Internet for all patrons was a clumsy way of protecting employees and children from material that might be found objectionable, the court held, the policy had to go.

It shouldn't be hard for judges to recognize public libraries as enclaves of free speech, since intellectual experimentation depends on the unrestricted flow of ideas. But the same Internet filtering policies that the courts have struck down in the context of libraries are increasingly common in college and universities. "By simply 'surfing' the net employees or students can be creating an uncomfortable and hostile environment for their colleagues," notes a recent article directed at employment lawyers. "Although a school by its very nature must provide for the guarantees of free speech as to classroom expression and assignment, the use of computers, [including] access to the Internet in open computer labs, should be appropriately regulated to avoid a hostile environment for offended students."[65]

In 1998, the same Virginia judge who invalidated the Loudoun County Public Library's filtering policy also struck down a Virginia law that restricted the ability of state employees to "download, print, or store" any sexually explicit material on state-owned computers. The law, purportedly designed to prevent sexual harassment, was challenged by professors at various

Virginia state colleges and universities, who said that it interfered with their ability to do research. (One professor complained that his Web site on gender roles and sexuality had been
censored; another was concerned about his ability to access Virginia's own database of sexually explicit poetry to continue his
studies on the "fleshy school" of Victorian poets.) In striking
down the law as a clear violation of academic freedom, the
judge noted that it seemed designed to discourage discourse
about sexual topics "not because it hampers public functions
but simply because [the state] disagree[s] with the content of employees' speech."[66]

Other Internet filtering policies at state universities have been
upheld, however,[67] and in 1999 the Supreme Court held, in a
5–4 decision, that even sexual teasing among elementary school
and college students can be a form of actionable gender discrimination. Federally funded schools and universities, the
Court held, may be liable for the sexual behavior and speech of
their students whenever they act with "deliberate indifference"
to harassment that is "so severe, pervasive and objectively offensive that it effectively bars the victim's access to an educational opportunity." In her majority decision, Justice O'Connor
stressed that harassment had to be offensive enough to change
the educational environment, and that schools can avoid liability if their response to known acts of harassment isn't "clearly
unreasonable."[68]

But the dissenting justices were unconvinced. Sexual harassment law is so vague, they recognized, that it can encompass any
sexually related behavior or speech, from practical jokes to
childish advances, that a reasonable person might find hostile or
offensive. In an effort to avoid liability, prudent school administrators will have an incentive to monitor and punish far more

sexual expression than the law actually forbids. Writing for the dissenters, Justice Kennedy said that the ambiguity of harassment law would force school administrators to operate in a "climate of fear." For example, he wrote, a female student who alleged "that a boy called her offensive names . . . that she told a teacher, that the teacher's response was unreasonable, and that her school performance suffered as a result, appears to state a successful claim."

In colleges and universities, the threats to the privacy of students may be even more acute. Because schools will be responsible for responding to "conduct that occurs even in student dormitory rooms," Kennedy observed, "schools may well be forced to apply workplace norms in the most private of domains."[69] The result might be an increased use of speech codes, he warned, to ensure that one student's dirty jokes don't become another student's hostile environment. But because speech codes raise serious First Amendment concerns, schools and universities may be put on the razor's edge of liability, as they risk being sued if they monitor their students' private lives and also sued if they don't. If Kennedy's fears are vindicated— and Justice O'Connor tried to assuage them in her opinion for the Court[70]—then colleges may be forced to monitor the Internet browsing habits and e-mail of students in dorm rooms, which seems hard to reconcile with the spirit of intellectual freedom that universities exist to preserve.

Harassment law now seems to be evolving in the direction that Kennedy feared. In 1999 the Vermont Human Rights Commission concluded that a Goddard College employee had created a hostile environment when he sent himself a sexually explicit greeting card over a university network, using the name of Katherine Kavanagh, a former Goddard College graduate student. While she was at home, Kavanagh read a message confirming that the e-mail had been sent, but the Vermont Human Rights Commission concluded that the Goddard College Internet Service Provider was a place of "public accommodations" and that Goddard College therefore had a legal obligation to prevent and eliminate any harassing speech that was transmitted over its network, even if the messages were read in the privacy of someone's home.[71] If this reasoning becomes more widely accepted, then even AOL would have a legal responsibility to monitor the content of all e-mail sent over its servers, on the grounds that cyberspace itself is a place of public accommodation.

I do not mean to suggest that employer monitoring of e-mail and Internet use is always inappropriate, or that the Supreme Court is responsible for all of the excesses that are committed in its name. But the e-mail and Internet monitoring that harassment law encourages inevitably leads to misunderstandings as jokes and comments intended for one audience are revealed to another. In this regard, Lawrence Lessig, Magno Ortega, and the Nissan employees are in the same boat: all experienced the invasion of privacy that results when private comments intended for one audience are exposed to another.

There is no reason for us to accept the passive view that all e-mail messages sent over a university or corporate network must be considered public rather than private. We should try instead to carve out backstage areas where people can joke, let down their hair, and form intimate relationships free from official scrutiny. Some studies suggest that 40 percent of e-mail correspondence is unrelated to work;[72] rather than trying in vain to exert managerial control over a medium that lends itself naturally to informal, and even irresponsible, communication, companies and universities might instead set aside e-mail accounts and areas of their networks as private places where speech and communications cannot be invaded unless there is cause to suspect individual users of serious misconduct. Unfortunately, this effort to distinguish between public and private speech in the workplace is something that current law refuses to allow. And the pressure on employers to monitor their employees' Internet use and e-mail threatens to subvert the Internet's greatest strength, which is to increase privacy and free expression by giving individuals more autonomy to decide how much of themselves to disclose to others. The challenge of refining the substantive definition of harassment so that the law protects privacy rather than threatens it is the subject of the next chapter.

Chapter 3

Jurisprurience

In 1997, Bob Guccione Jr., the publisher of *Spin* magazine, was put on trial for sexual favoritism. A former editor named Staci Bonner claimed that she hadn't been promoted to a "creative" position because Guccione favored "tall, thin blondes" in the workplace who slept with him or flirted with him, while passing over women to whom he wasn't attracted. The principal victim of the lawsuit, however, wasn't Guccione, but instead a group of women in the workplace whom Bonner believed had benefited from having consensual affairs with Guccione and his male editors. Most aggrieved of all was Guccione's former girlfriend, Celia Farber, who wrote a column about AIDS for *Spin,* and who had been involved, for a few years, in a serious relationship with her mentor and boss. In an article for the on-line magazine *Salon,* Farber described her bewilderment at a suit that struck her and her colleagues as implausible and destructive. "Whose lunatic idea it was I don't know, but they actually set out to explain every last byline, promotion and even desk placement at a struggling rock magazine according to the sexual dynamics of the workplace." The result was that dozens of women's private lives were "ransacked, scrutinized and judged."[1]

Farber said she was humiliated at having the most intimate details of her love life exposed to public view. "Over a period of almost three years, I was subpoenaed for thousands of pages of documentation, including every draft of every article I ever wrote, my entire academic record, love letters, diaries and documentation of any 'dates,' 'gifts' or any other physical remains of the affair," she wrote. "I was a witness, but treated with the kind of hostility usually reserved for a defendant." Farber lost ten pounds and borrowed money to pay her legal bills. Her professional achievements were belittled, as Bonner's lawyer tried to suggest that she had been rewarded because of her affair, even though she had held the same job for ten years, and her salary had increased more slowly than that of her colleagues. Other employees suffered similar ordeals. Minor failings, like sloppy photocopies, were introduced as evidence of the incompetence of "favored" women. "Witnesses, mostly former or present *Spin* employees, were brought in from as far away as Taiwan to comb their memories of every expletive shouted by an editor on deadline, every crude remark, every interoffice liaison, every flirtation," Farber wrote. "One woman, who had dated Guccione briefly years after the plaintiff left the magazine, was forced, during a deposition, to graphically detail a sexual experience with him to a thundering (male) attorney—who was ostensibly fighting on the side that was against sexual harassment."[2]

After a four-week trial, a jury found that Bonner had suffered from a hostile working environment, created by editors who were prone to sexist insults such as "girls can't write" and who commented on women's appearance in the most vulgar terms. But the jury found no evidence that Bonner had been the victim either of sexual favoritism or of quid pro quo sexual harassment. Bonner had presented "ample evidence" that "there was a system of sexual favoritism at *Spin* whereby a female employee could not advance in the editorial department unless she submitted to the sexual advances of Mr. Guccione or other senior male editors," Judge Denise Cote observed in explaining the jury's verdict. But Bonner had not presented convincing evidence that "she herself was asked to submit to sexual advances in order to gain a promotion" or that "she was sufficiently qualified for the positions to which she aspired."[3] Farber, for her part, remains outraged by the notion that a colleague, simply by alleging sexual favoritism to explain her own lack of promotion, could turn an entire workplace upside down, invading the privacy of innocent women in the process. "I am not attempting to whitewash *Spin*," Farber wrote. "It's just that rude behavior does not belong in federal court. Real sexual harassment does. Harassment where sex is used to control women in the workplace. I know of no shred of evidence in this entire case that suggests such a quid pro quo dynamic."

The *Spin* case was a harbinger of things to come. A few months after the Guccione verdict, Paula Jones amended the complaint in her sexual harassment suit against President Clinton. Jones, too, justified her efforts to pry into Clinton's consensual relationships, and to ask him to name all the women with whom he had proposed to have sexual relations as governor and president, on the grounds that Clinton had engaged in illegal sex discrimination when he favored employees who succumbed to his advances and punished those who did not. And Lewinsky, like Celia Farber, was outraged that another woman's right to pursue an implausible accusation of sexual favoritism took precedence over her own right to privacy.

The Bonner and Jones suits suggest that something has gone wrong in the law of sexual harassment. A legal cause of action originally designed to protect the privacy and dignity of women and men in the workplace has, on several occasions, permitted unreasonable invasions of the privacy and dignity of women and men in the workplace. "The cure," as Farber puts it, "is fast becoming worse than the disease." If harassment law did nothing more than threaten the privacy of innocent third parties, such as Celia Farber and Monica Lewinsky, it might be argued that their privacy claims should be balanced against the privacy claims of women like Staci Bonner and Paula Jones. But the experiences of Lewinsky and Farber are symptoms of a broader problem with harassment law, which threatens not only the privacy of innocent bystanders but also the boundaries between the private and public spheres.

In this chapter, I would like to examine the sources of this threat to privacy, and to propose a way of resurrecting some of the privacy we have lost. The solution, I will suggest, lies in returning to the text of the Civil Rights Act of 1964, which properly ensures that no American should have decisions about his or her livelihood turn on his or her sexuality. Any speech or conduct that limits an employee's professional status or opportunities, or affects the terms and conditions of employment, because of sex—including not only explicit and implicit sexual threats but also employment decisions based on stereotypical views of the individual's abilities—is and clearly should be forbidden by law. At the same time, I will suggest, speech and conduct that other employees may find offensive, but that are not severe or pervasive enough to change the terms and conditions of employment, should be regulated by invasion of privacy law rather than by employment discrimination law. Whenever the conduct in question can't be viewed as a clear case of gender discrimination—in cases involving consensual affairs, offensive jokes, and vulgar e-mail—I will argue that invasion of privacy law is better suited than the Supreme Court's "hostile environment" test to distinguish between relatively trivial indignities and those that are serious enough to be illegal.

To understand how far harassment law has strayed from its roots, let's begin with the text of Title VII of the Civil Rights Act of 1964, which says:

> It shall be an unlawful employment practice for an employer . . . to fail or refuse to hire or to discharge any individual, or otherwise to discriminate against any individual with respect to his compensation, terms, conditions, or privileges of employment, because of such individual's race, color, religion, sex, or national origin; or to limit, segregate, or classify his employees or applicants for employment in any way which would deprive or tend to deprive any individual employee of employment opportunities or otherwise adversely affect his status as an employee, because of such individual's race, color, religion, sex, or national origin.[4]

The word "sex" was added to the Civil Rights Act a day before its passage by Representative Howard Smith, a Southern Congressman and passionate segregationist who opposed the bill and sought to scuttle its passage. By all accounts, Smith viewed his proposal as something of a joke.[5] But what, precisely, does it mean to discriminate against any individual because of his or her sex? Here are three examples to test our intuition.

In 1990, after having an affair with him, Gennifer Flowers told Bill Clinton, then governor of Arkansas, that she wanted a state job. According to a declaration that Flowers filed in the Paula Jones case, Clinton suggested that Flowers contact his assistant, who would help her with the job application. As a result of the meeting, Flowers was eventually hired as an administrative assistant for the Arkansas Appeal Tribunal, the agency responsible for overseeing unemployment claims disputes. Soon afterward, an African-American woman named Charlotte Perry, who had applied for the same job, complained that she had been passed over because Flowers was having an affair with the Governor. A state grievance officer initially ruled in Perry's favor, finding that state officials had added a background in "public relations" as a new requirement for the job in order to suit Flowers's limited credentials. But the grievance officer was overruled by the head of the state agency that hired Flowers.[6] Was Perry a victim of discrimination because of her sex?

In 1988, Bill Gates, the chairman of Microsoft and the richest man in the world, met one of his junior employees, Melinda French, at a company picnic. According to the *Ladies' Home Journal,* French was openly affectionate with Gates from the start, but they dated for five years before he finally proposed.[7] Melinda Gates left Microsoft when her daughter Jennifer was born in 1996, but before her departure, during five years of courtship and three years of marriage to the CEO, she was a project manager in charge of overseeing multimedia and other software. Imagine that a group of male Microsoft employees sue the company, claiming that Melinda Gates was given special treatment in the workplace because of her relationship with the boss. The men claim that they might have had access to similar advantages but for the fact that they were men. Were they victims of discrimination because of sex?

In 1982, seven male respiratory therapists in Westchester County, New York, sued the hospital that employed them. They argued that their boss, James Ryan, had unfairly denied them the possibility of a promotion by requiring all applicants for the position of assistant chief respiratory therapist to be registered by the National Board of Respiratory Therapists. Ryan invented this requirement, they argued, in order to disqualify them and hire Jean Guagenti, a woman with whom he was having an affair. By favoring his female lover, the male therapists argued, Ryan had discriminated against them because of sex.

The Gennifer Flowers case was cut short, and the Melinda Gates case is hypothetical, but the case involving Jean Guagenti was decided in 1986 by the U.S. Court of Appeals in New York, when Judge Roger J. Miner held that Title VII doesn't forbid isolated cases of consensual sexual favoritism. The male employees "were not prejudiced because of their status as males; rather, they were discriminated against because Ryan preferred his paramour," Miner held. "Appellees faced exactly the same predicament as that faced by any woman applicant for the promotion: No one but Guagenti could be considered for the appointment because of Guagenti's special relationship with Ryan."[8] Miner was especially concerned about the prospect of involving federal courts in the "policing of intimate relationships": if employees were allowed to challenge the consensual affairs of their colleagues as a form of gender discrimination, he reasoned, then a complaint by a disappointed coworker would be enough to trigger a wide-ranging investigation into the private lives of consenting adults.

Judge Miner was surely correct to reject the argument that isolated consensual affairs are a form of sex discrimination. Some bosses may be more likely to trust people with whom they're romantically involved. If there's something unfair about sexual favoritism on the job, it has more to do with our intuitions about nepotism than with sex discrimination, and nepotism generally isn't illegal. Moreover, to allow jealous colleagues to make a federal case out of sexual favoritism would dramatically increase the role of federal courts in regulating the private lives of American workers. A decade ago, a study by the Bureau of National Affairs estimated that almost a third of all romances begin at work.[9] Around the same time, in a study of randomly chosen executive women, more than half the respondents said they were romantically involved with men they met through work.[10] If each of these romances could be challenged by coworkers, then the private lives of millions of innocent Americans would be vulnerable to judicial scrutiny.

The handful of cases that have recognized sexual favoritism as a form of sex discrimination have vindicated Judge Miner's fears about invasions of privacy. In the leading case, a female nurse at the District of Columbia jail argued that she was the victim of sex discrimination when she was denied a promotion that went to another nurse in her unit who was having an affair with the chief medical officer. In 1984, a federal district court agreed that sexual favoritism fit within the then existing law of sexual harassment. But it ruled against Ms. King anyway, on the grounds that she had failed to provide direct evidence of an explicit sexual relationship between the promoted nurse and the medical officer, both of whom denied that they were involved with each other. King had offered plenty of indirect evidence, including testimony that the suspected lovers attended out-of-town conventions together, took long lunches together, and displayed a "casual physical friendliness" at the office. The court said this was outweighed by evidence that the medical officer had a "settled" family life and the promoted nurse had an active sex life with other men. Claims of sexual favoritism, the court held, "must not rest on rumor, knowing winks and prurient overtones or on inferences allowed in divorce law."[11]

But the appellate court disagreed, refusing to put the plaintiff in the position of Iago, who accused Desdemona of committing adultery, only to have Othello demand "ocular proof." Circumstantial evidence of sexual conduct, such as "kisses, embraces and other amorous behavior" would do.[12] The appellate court dismissed the concern that it was opening the floodgate to jarring invasions of privacy—allowing jealous coworkers to charge, for example, that a candidate selected for promotion was sexually "attractive" to the selecting officials. But, as the *Spin* case shows, it's hard to avoid asking whether an employer's sexual attraction led to favors for a particular employee, if the test for harassment is whether an employee would have been treated differently if he or she were a member of the opposite sex.

When it came time for the Equal Employment Opportunity Commission to formulate a policy about sexual favoritism in 1990, it rejected this approach.

It is the Commission's position that Title VII does not prohibit isolated instances of preferential treatment based upon consensual romantic relationships. An isolated instance of favoritism toward a "paramour" (or a spouse, or a friend) may be unfair, but it does not discriminate against women or men in violation of Title VII, since both are disadvantaged for reasons other than their genders,[13]

declared the EEOC Policy Guidance on Sexual Favoritism, formulated under the direction of Clarence Thomas.[14] But the EEOC then confronted a dilemma that reveals the analytical confusion at the heart of sexual harassment law. In the 1980s, the EEOC proposed, and the Supreme Court agreed, that illegal sexual harassment might be inferred whenever the plaintiff would have been treated the same as all other workers if he or she were a member of the opposite sex. The same legal analysis that courts had rejected in the sexual favoritism cases had, by 1990, become the foundation of the EEOC's own definition of illegal sexual harassment.

In an attempt to reconcile its conflicting impulses, the EEOC recognized two forms of sexual favoritism that might be illegal. First, the EEOC announced that "[i]f a female employee is coerced into submitting to unwelcome sexual advances in return for a job benefit, other female employees who were qualified for but were denied the benefit may be able to establish that sex was generally made a condition for receiving the benefit."[15] Second, the commission held that "[i]f favoritism based upon the granting of sexual favors is widespread in a workplace, both male and female colleagues who do not welcome this conduct can establish a hostile environment in violation of Title VII regardless of whether any objectionable conduct is directed at them and regardless of whether those who were granted a favorable treatment willingly bestowed the sexual favors. In these circumstances, a message is implicitly conveyed that the managers view women as 'sexual playthings,' thereby creating an atmosphere that is demeaning to women."[16]

The EEOC's effort to permit isolated affairs at work, but to prohibit a widespread pattern of sexual favoritism, makes some intuitive sense. If consensual affairs between supervisors and subordinates are so widespread that employees who *don't* sleep with the boss can't get ahead, then sexual compliance has become a "term or condition" of employment—precisely what Title VII was intended to prohibit. The classic case of harassment involves a supervisor who threatens to retaliate against women who don't succumb to his sexual demands. This is what is known as quid pro quo harassment: sleep with me or you're fired. In cases where sexual favoritism pervades the workplace, then employees might reasonably fear that sexual compliance has become an implicit rather than explicit term or condition of employment, and the violation of Title VII is similarly clear.

But the case of a lone supervisor who is a serial monogamist, and who engages in a series of consensual but ultimately unsuccessful affairs with women in the workplace, is a harder question. On the one hand, a pattern of having affairs with subordinates, even if the affairs weren't coercive, might undermine the professionalism of the workplace, discomfiting male and female employees, who might legitimately worry that they were being passed over for job opportunities that their more complaisant colleagues received.[17] On the other hand, reasonable employees might disagree about the message conveyed by a supervisor who engages in a series of isolated consensual affairs. Far from viewing women as "sexual playthings," the supervisor might be attracted to particular women in the workplace precisely because of their intelligence and professional competence.

In cases where it's possible to disagree about whether a particular pattern of conduct in the workplace tends to "discriminate against any individual with respect to his [or her] compensation, terms, conditions, or privileges of employment, because of . . . sex," or whether an employer is classifying employees in a way that tends to deprive any individual of "employment opportunities or otherwise adversely affect his [or her] status as an employee, because of . . . sex," my own instinct is that the employment discrimination law has no business regulating the conduct in question.

As it has evolved, however, sexual harassment law is ill equipped to give a clear answer to the question of whether even isolated consensual affairs are or are not a form of illegal sex discrimination. And this ambiguity stems from the intellectual roots of harassment law in a strain of feminist theory that stressed the connection between male desire and male dominance. In the 1960s and 1970s, early theorists of sexual harassment insisted that heterosexual sexual relations and sexual domination were inextricably linked, and that expressions of male heterosexual desire themselves tended to subordinate women. In her book *Sexual Shakedown,* published in 1978, Lin Farley defined harassment as "unsolicited nonreciprocal male behavior that asserts a woman's sex role over her function as a worker," and she suggested that any conduct or speech in the workplace that defined women in sexual terms might be viewed as a form of sex discrimination. Harassment, Farley said, can be "any or all of the following: staring at, commenting upon, or touching a woman's body; requests for acquiescence in sexual behavior, repeated nonreciprocated propositions for dates; demands for sexual intercourse; and rape."[18]

A year later, Catharine MacKinnon developed the argument

that gender inequality is inextricably linked with heterosexual desire. In *Sexual Harassment of Working Women: A Case of Sex Discrimination*,[19] MacKinnon insisted that some sexual expression might be viewed as inherently discriminatory, because heterosexual relations were the means by which male domination and female subordination were enforced. In MacKinnon's words, "practices which express and reinforce the social inequality of women to men are clear cases of sex-based discrimination in the inequality approach."[20] She began her book by defining sexual harassment as "the unwanted imposition of sexual requirements in the context of a relationship of unequal power."[21] But the innovation of her book was to identify a new category of sexual harassment that didn't require evidence of physical contact, explicit threats, or tangible job consequences to be illegal. She described this as the situation where sexualized comments or displays in the workplace made the "work environment unbearable." Such a hostile environment was discrimination "based on sex," according to the inequality approach, whenever "women are socially defined as women largely in sexual terms."[22]

Although MacKinnon was vague about what conduct her vision of sexual harassment would prohibit, the logical implications of her inequality approach seemed to call into question a great deal of sexual expression in the workplace. For male sexuality itself, MacKinnon famously argued, is the lynchpin of gender inequality. "The male sexual role . . . centers on aggressive intrusion on those with less power. Such acts of dominance are experienced as sexually arousing, as sex itself,"[23] she wrote. "A feminist theory of sexuality," therefore, "locates sexuality within a theory of gender inequality, meaning the social hierarchy of men over women."[24] Anatomy is destiny, or rather law.

In her less guarded rhetoric, MacKinnon seemed to suggest that even apparently consensual heterosexual sex might reinforce the social inequality of women to men, and thus constitute sex discrimination, at least when it takes place between men and women whose social status is not precisely equal. "[I]s ordinary sexuality, under conditions of gender inequality, to be presumed healthy?" MacKinnon asked in a crucial passage.

> What if inequality is built into the social conceptions of male and female sexuality, of masculinity and femininity, of sexiness and heterosexual attractiveness? Incidents of sexual harassment suggest that male sexual desire itself may be aroused by female vulnerability. . . . Analysis of sexuality must not be severed and abstracted from analysis of gender. . . . If sexuality is set apart from gender, it will be a law unto itself.[25]

In later works, MacKinnon explained that her arguments "that defined sexual harassment as a form of discrimination on the basis of sex" are based on "the same view of sex discrimination that underlies"[26] her efforts to ban pornography as an expression of female subordination. The root meaning of "pornography," MacKinnon and Andrea Dworkin noted in a memo about their bill to ban pornography as sex discrimination in Minneapolis in 1983, is "the graphic depiction of whores. . . . In pornography, women are graphically depicted as whores by nature, that is, defined by our status as sexual chattel."[27] It is the depiction of women in subordinate positions, MacKinnon argues, that makes pornography and harassment forms of sex discrimination: "pornography as we define it makes the inequality of the sexes sexual, the way that it makes it sexy, the way that it eroticizes putting women in an inferior position."[28]

Courts and citizens have had no trouble understanding that MacKinnon's efforts to regulate pornography as a form of sex discrimination violate the core protections of the First Amendment. Her Minneapolis ordinance defined pornography as "the sexually explicit subordination of women, graphically depicted, whether in pictures or in words, that also includes one or more of the following: women are presented dehumanized as sexual objects, things or commodities . . . women are presented in postures of sexual submission," where "women are presented as whores by nature."[29] "This is thought control,"[30] Judge Frank Easterbrook noted in his opinion striking down a similar ordinance that MacKinnon had sponsored in Indianapolis in 1984:

> Speech treating women in the approved way—in sexual encounters "premised on equality"—is lawful no matter how sexually explicit. Speech treating women in the disapproved way—as submissive in matters sexual or as enjoying humiliation—is unlawful no matter how significant the literary, artistic or political qualities of the work taken as a whole. The state may not ordain preferred viewpoints in this way.[31]

What is surprising is that judges and citizens have been so much slower to grasp that MacKinnon's view of sexual harassment threatens values of free expression for precisely the same reason that her vision of pornography does. A year after the publication of her book, the Equal Employment Opportunity Commission issued guidelines declaring that sexual harassment should be defined to include not only quid pro quo threats but also "verbal or physical conduct of a sexual nature" that has "the purpose or effect of unreasonably interfering with an individual's work performance or creating an intimidating, hostile, or offensive working environment."[32] And in 1986, in the land-mark Meritor case, the Supreme Court unanimously endorsed the EEOC's "hostile environment" test.

In 1974, Mechelle Vinson was hired to work as a teller-trainee at the Meritor Savings Bank under the supervision of Sidney Taylor, the branch manager. Vinson claimed that Taylor invited her out to dinner soon after the end of her probationary period, and during the meal he suggested that they go to a motel to have sex. She initially refused, but eventually agreed, out of fear of losing her job. Taylor continued to make demands on her for sexual favors, and Taylor estimated that over the next several years she had sex with him forty or fifty times. She also claimed that Taylor fondled her in front of other employees, exposed himself to her in the women's rest room, and forcibly raped her. During this period, she advanced from teller to assistant branch manager and never complained about Taylor's behavior. But after four years, Vinson took a sick leave, and when she stayed away for two months, the bank fired her for excessive use of her leave. She then sued the bank, on the grounds that Taylor had subjected her to sexual harassment during her four years on the job.

The Supreme Court might have held that Taylor's constant sexual demands on Vinson clearly changed the terms and conditions of her employment because of sex—an obvious violation of the text of Title VII. Instead, Chief Justice William H. Rehnquist departed from the text of Title VII, holding that Taylor had discriminated against Vinson by creating a sexually charged hostile and offensive working environment. "Without question, when a supervisor sexually harasses a subordinate because of the subordinate's sex, that supervisor 'discriminate[s]' on the basis of sex," Rehnquist wrote in a crucial sentence that begged the central question the Court had been asked to resolve. Rehnquist did not provide a clear definition of what, precisely, hostile environment harassment was, or why it should viewed as a form of sex discrimination, except to say that "for sexual harassment to be actionable, it must be sufficiently severe or pervasive to alter the conditions of [the victim's] employment and create an abusive working environment."

The hostile environment test transformed the nature of the workplace by blurring the boundaries of the public and private spheres. By allowing women (or men) to complain about any sexually oriented speech or conduct that they found hostile or abusive, the new test allowed aggrieved coworkers to object to overheard jokes and to e-mail, suggestive pictures, or even their colleagues' consensual flirtation, even if the men in question never intended their conduct to be offensive, and the women to whom the conduct was directed didn't perceive it as offensive. And unlike quid pro quo harassment, a hostile environment doesn't require any proof of economic or even psychological injury. While the class of women seeking to recover for quid pro quo harassment was limited to those who suffered tangible harm, under the hostile environment theory of harassment, the "hostility" or "offensiveness" of the speech became the test of whether or not, in legal terms, harm had been suffered.

Throughout the 1980s and early 1990s, critics of the hostile environment test argued that it posed a serious threat to privacy and free expression. In 1992, an article by Eugene Volokh of the UCLA Law School collected some of the most extreme examples. One court said that the use of gender-based job titles (like "draftsman" or "foreman") could be actionable. Another court required an employer to proscribe any "sexually suggestive" material, defined broadly enough to include reproductions of classical paintings. Volokh and other critics of harassment law argued that the only way to tie harassment law to the text of Title VII, which prohibits gender discrimination "against any *individual,*" would be to require that speech be directed at a particular employee in order to be actionable, rather than offending women in general. A group called Feminists for Free Expression argued that the hostile environment test subverted equality as well as free speech in the workplace. The effort to shield women by bowdlerizing the speech of men enshrined archaic stereotypes of women as delicate creatures who need special protection from words and images, these feminists argued, and the generalization that women are more likely to be offended by scatology than men seemed like precisely the kind of romantic paternalism that the Civil Rights Act was designed to erase. Accordingly, the Feminists for Free Expression proposed the following replacement for the hostile environment test: a pattern of conduct or expression which is directed at a specific employee; which a reasonable person would experience as harassment; and which substantially hindered the employee's job performance.

Unfortunately, the Supreme Court was not interested in working out the analytical and practical details of the legal revolution it had set in motion. In a 1993 case, *Harris v. Forklift Systems,* the Court blandly reaffirmed the hostile environment test without acknowledging any of the confusion it had provoked. Even if it had no tangible effects on a complainant's job performance, Justice Sandra Day O'Connor suggested for the Court, sexually oriented expression in the workplace might qualify as sexual harassment if it was severe or pervasive enough "to create an objectively hostile or abusive work environment—an environment that a reasonable person would find hostile or abusive," in addition to having been perceived as hostile or abusive by the plaintiff herself. But O'Connor provided no useful guidelines for defining whether or not conduct was "discriminatory," "severe or pervasive," or "hostile or abusive," or how to measure the sensibility of a "reasonable person." She noted unhelpfully that "whether an environment is 'hostile' or 'abusive' can be determined only by looking at all the circumstances." In a concurring opinion, Justice Antonin Scalia worried that "as a practical matter, today's holding lets virtually unguided juries decide whether sex-related conduct engaged in (or permitted by) an employer is egregious enough to warrant an award of damages." But he then threw up his hands, saying, "I know of no test more faithful to the inherently vague statutory language than the one the Court today adopts."

Although the Supreme Court announced that the relevant question was whether a "reasonable person" would find the conduct in question "hostile or abusive," lower courts have continued to grapple with the ambiguities that the justices refused to resolve. Some courts ask whether a "reasonable woman" would find the conduct offensive, reasoning that a "sex-blind reasonable person standard tends to be male-biased and tends to systematically ignore the experiences of women."[33] But the reasonable woman standard runs the risk of holding men responsible for well-intentioned advances that they didn't expect would be perceived as offensive. Even more troubling, it promotes the sexist assumption that gender discrimination law is designed to erase—namely, that women need special treatment, rather than equal treatment, in the workplace. Moreover, the Supreme Court has traditionally protected offensive speech because "one man's vulgarity is another's lyric." Under the gender-specific standard, speech can be banned whenever one man's lyric becomes a reasonable woman's vulgarity. The fact that different women find different things offensive makes the reasonable woman standard even more confusing. Some courts have tried to avoid the problem by examining the perspective of a "reasonable person in the victim's position." But this highly subjective test only compounds the difficulties of the reasonable woman standard, requiring judges to imagine the reaction of a reasonable African-American women, or a reasonable Catholic woman, and so on, depending on the identity of the victim.

In an attempt to reconcile harassment law with gender-blind ideals, some courts have tried to conceive of the "reasonable person" as a sex-blind abstraction that incorporates the perspectives of men and women. But empirical evidence suggests, unsurprisingly, that men and women perceive the same sexual encounters very differently: men are more likely to be flattered or amused by sexual displays at work, while women are more likely to feel insulted or frightened.[34] The reasonable person standard gives juries little guidance about how to reconcile perspectives that are, at times, irreconcilable. Are people from Mars more reasonable than people from Venus?

Because ambiguous legal standards have chilling effects, prudent companies have felt increasing pressure to ban more speech than the law actually prohibits. Most courts that have considered the issue, including Judge Susan Webber Wright in the Paula Jones case, have assumed that any sexual advance that a reasonable woman would find offensive can constitute intentional discrimination "on the basis of sex," even if the advance was well intentioned and the offense was unintentional. But this has opened the floodgates to the policing of consensual affairs in precisely the way that the Chicago court feared: there is evidence suggesting, for example, that more than a quarter of all sexual harassment claims today originate from soured romances.[35]

Moreover, the effort to distinguish welcome from unwelcome advances has led courts to enumerate, with the prim logic of a Victorian courtship manual, the full range of possible responses to a sexual advance. As the leading casebook on the subject puts it: "The complainant's possible responses include (1) outright rejection, (2) initial rejection and later acceptance, (3) initial acceptance followed by later rejection, (4) ambiguous conduct, (5) coerced submission, and (6) welcome acceptance."[36] In each of these categories, the problems of proof are daunting, and inevitably involve invasions of the privacy of both plaintiff and defendant, exposing to public view a great deal of private conduct that is ultimately found to be innocuous.

How could sexual harassment law be refined so that it protects privacy instead of threatening it? In my view, the hostile environment test has proved more distracting than clarifying in identifying illegal gender discrimination, and it should be eliminated. Instead, harassment law should be reoriented toward the text of Title VII of the 1964 Civil Rights Act. Judges and juries in harassment cases should focus on this question: can the disputed speech or conduct be viewed as a form of discrimination "because of sex"? Indeed, the Supreme Court in recent cases has been moving in this direction, minimizing the distinction between quid pro quo and hostile environment harassment. As Justice Ginsburg put it in the Harris case, "The critical issue, Title VII's text indicates, is whether members of one sex are exposed to disadvantageous terms or conditions of employment to which members of the other sex are not exposed."[37] Let's try to imagine what the world would look like if the hostile environment test were abandoned.

The existing law of quid pro quo harassment would be unaffected by the repudiation of the hostile environment test. Title

VII makes it illegal "to discharge any individual, or otherwise to discriminate against any individual with respect to his compensation, terms, conditions, or privileges of employment, because of an individual's sex." With the help of the crucial ellipsis, it's not hard to conclude that implicit or explicit threats to fire someone for failure to submit to sexual demands is a form of discrimination in the terms or conditions of employment "because of . . . sex" (although not necessarily "because of an *individual's* sex"—in other words, not necessarily because of gender).

Similarly, any evidence of pervasive gender-based animosity, or of differential treatment of women based on stereotypical views about their abilities, would change the terms and the conditions of employment in a way that would obviously constitute discrimination because of sex. The often overlooked second sentence of the relevant section of Title VII makes it illegal to "limit, segregate, or classify" employees in a way that tends to "deprive any individual employee of employment opportunities or otherwise adversely affect his [or her] status as an employee because of . . . sex." This covers the gauntlet situations that seem to be more common in blue-collar professions, in which women are forced to endure a pattern of sexist taunts and abuse on a daily basis.[38] It also covers the subtler patterns of sexist treatment chronicled by Vicki Schultz of Yale Law School, in which the professional opportunities of women are limited, and their professional abilities are belittled, because of the discriminatory attitudes of their male employers.[39]

That leaves only one category of sexual harassment that might be less closely regulated by Title VII if the hostile environment test were abandoned: speech and conduct of a sexual nature that can't plausibly be characterized as sex discrimination.

This might include unwanted advances, suggestive looks and gestures, sexual joking and teasing, and the display of sexually explicit material that carry no implicit or explicit threat of retaliation, don't change the terms and conditions of employment, and don't deprive any individual of employment opportunities or adversely affect his or her status as an employee because of sex, but are nevertheless annoying and offensive.

There is room for vigorous debate about precisely how much speech and conduct this category includes. Sexual harassment is highly dependent on context: a series of unwanted requests for dates from a supervisor might well be viewed as changing the terms and conditions of employment in a discriminatory way, while a series of unwanted requests for dates from a coworker might not. Similarly, a pattern of sexist taunts that a supervisor directs at a disfavored worker might well affect her status as an employee, while overheard vulgar jokes that two male colleagues make about the same woman might not. But if the hostile environment test were abandoned, this debate about whether speech and conduct are serious enough to change the terms and conditions of employment, and are thus properly considered evidence of gender discrimination, could be guided by the text of Title VII itself.

Assume, for example, that Governor Bill Clinton did indeed expose himself to Paula Jones, and then retreated after being rebuffed. Some people might agree with Jones that Clinton discriminated against her, and affected her employment status, by conveying the impression that he saw her as a sex object rather than a professional. Others might conclude that because Clinton had only nominal supervisory authority over Jones, and because the advance took place outside the workplace, it should be considered a private rather than an official act. Employment discrimination law, I think, should be agnostic between these two positions, and intervene only when the connection between the conduct and the conditions of employment is clear.

Or consider Monica Lewinsky's plight. Jones's lawyers argued repeatedly that Clinton's relationship with Lewinsky could not have been consensual, because of the unequal power relationship between them. John Whitehead, the president of the Rutherford Institute, which paid Paula Jones's legal expenses, is a born-again Christian who admires the work of Catharine MacKinnon and adorns his office with autographed posters of work by the feminist artist Judy Chicago, which features female genitalia served up on dinner plates. "That's the connection between Lewinsky and Jones," he told me. "Clinton's taking advantage of people." Lewinsky herself strenuously resisted efforts to portray herself as a victim, however, and after the facts emerged, it was difficult to view Clinton as Lewinsky's oppressor. "She wanted to have sex with him at least as much as he wanted to have sex with her, and probably more," Judge Richard Posner notes. "Their relationship involved virtually a reversal of the traditional sex roles."[40] Because people can disagree about whether Lewinsky's relationship with Clinton was or was not coercive, it should not be considered evidence of gender discrimination.

A new generation of feminist scholars is beginning to recognize the analytical awkwardness of defining sexual harassment exclusively as a form of gender discrimination. In an article in the *Harvard Law Review* in 1997, Anita Bernstein argued that sexual harassment should be understood as a dignitary injury. The harms that hostile environment sexual harassment inflicts, Bernstein argued, are indignity and humiliation, and are therefore better understood not only as a form of gender discrimina-

tion but also as something more like what the law calls a tort—
a civil injury inflicted by one individual upon another.[41] Tort law
is the area of our legal system that provides remedies against
those who have transgressed social norms of decency or propriety. In my view, however, the indignity of hostile environment
harassment can often be defined more precisely, as an invasion
of privacy. Privacy, as I suggested in the Prologue, protects us
from being objectified and simplified and judged out of context
in a world of short attention spans, a world in which part of
our identity can be mistaken for the whole of our identity. Remember Milan Kundera's observation that when privacy is
invaded, the transformation of an individual from subject to
object is experienced as shame. Like hate speech, which stereotypes and oversimplifies an individual on the basis of a single
characteristic—such as race or sexual orientation—an unwanted
advance can objectify a woman on the basis of her appearance,
allowing one aspect of her identity to overwhelm all other aspects of her identity. In this sense, the indignity that Paula Jones
suffered if her willingness to flirt with Clinton was taken out of
context is similar to the indignity that Lawrence Lessig suffered
when his e-mail taken out of context. The humiliation that results from being objectified and stereotyped on the basis of an
isolated characteristic that comes to define one's public face is
the distinctive injury of an invasion of privacy.

Indeed, much of the noncoercive behavior that hostile environment law currently regulates might be reconceived as an
offense against privacy. It often involves a breakdown in the
communication of what Erving Goffman called "embodied
messages"[42]—the glances, gestures, and spoken words that men
and women use as measures of whether or not they want to be
accessible to each other in social situations. As the range of re-

sponses that courts have enumerated in trying to distinguish welcome from unwelcome advances suggests, the difficult cases arise in situations where the alleged harassers have misunderstood the way their behavior is being perceived and the alleged victims have failed to make their reactions entirely clear. In a case involving a situation where the alleged harasser was making what he believed to be "innocent or invited overtures," a judge on the U.S. Court of Appeals for the First Circuit recognized the complexity of social interactions, where both parties, in order to behave appropriately, must be sensitive to what are often extremely subtle cues:

> The man must be sensitive to signals from the woman that his comments are unwelcome; and the woman, conversely, must take responsibility for making those signals clear. In some instances, a woman may have the responsibility for telling the man directly that his comments or conduct is unwelcome. In other instances, however, a woman's consistent failure to respond to suggestive comments or gestures may be sufficient to communicate that the man's conduct is unwelcome.[43]

Under the guise of rooting out gender discrimination, courts today often find themselves policing privacy violations, as they are forced to decide whether or not a man behaved unreasonably in failing to perceive the woman's discomfort, or the woman behaved unreasonably in failing to make her discomfort clear. The privacy violation takes the form of invading the boundaries of a person who has indicated, directly or indirectly, that she or he wishes to remain inaccessible. But determining whether or not the failure of mutual respect should be illegal is no easy task. In cases involving an ambiguous response, courts sometimes reject harassment claims where employees, because of politeness, indecision, or fear, fail to indicate that their supervisors' attentions are unwelcome.[44] In a case involving mixed signals, where a complainant invited her supervisor to discuss promotion opportunities in a bar, courts have concluded that her behavior "induced" the sexual advance that followed.[45]

Although some forms of harassment are better understood as invasions of privacy rather than as gender discrimination, not all invasions of privacy should be regulated by law. In America, courts today are properly reluctant to restrict hate speech, pornography, seditious libel, and blasphemy—all of which are offenses against norms of dignity and propriety—because of their potential for inhibiting freedom of speech and thought. When citizens who have been objectified and misjudged seek legal remedies for their humiliation, there is a similarly high risk of thought control: the right to misjudge one another is a hallmark of citizens in a free society. In cases where harassment rises to the level of gender discrimination, courts can plausibly balance two competing constitutional values—free expression on the one hand and the right to be free from discrimination on the other. But when harassment doesn't rise to the level of discrimination, and is instead an offense against personal dignity—a right not guaranteed by the Constitution, at least not in the abstract—then it's arguable that the federal government has no compelling interest in regulating it. Instead, in cases where there are no tangible employment consequences, invasion of privacy law is better equipped than gender discrimination law to distinguish truly egregious violations, which should be illegal, from merely offensive and inappropriate behavior, which should not.

The most intentionally hurtful and harassing behavior can be regulated by the intentional torts: battery, assault, false imprisonment, and outrage. Unwanted groping sometimes rises to the level of assault and battery. As for the tort of outrage, the standard for liability is very high indeed: to be illegal, the conduct must be "outrageous in character and so extreme in degree as to go beyond all possible bounds of decency, and to be regarded as atrocious and utterly intolerable in a civilized society."[46] The accuser has to prove that the accused intended to inflict emotional distress, that the conduct was extreme and outrageous, and that the distress was severe. In the most extreme sexual harassment cases, involving repeated examples of deliberate and malicious harassing behavior, plaintiffs have sometimes prevailed under the tort of outrage.[47]

But many sexual harassment allegations fall short of conduct plainly intended to harm or humiliate, and look more like boorish advances that the plaintiff failed to make clear were unwelcome and that the defendant insensitively pressed despite the warnings he may have received. In this gray area, the torts of invasion of privacy are especially clarifying. Recall the definition of an intrusion on seclusion: "One who intentionally intrudes, physically or otherwise, upon the solitude or seclusion of another or his private affairs or concerns, is subject to liability to the other for invasion of his privacy, if the intrusion would be highly offensive to a reasonable person."[48]

The intrusion on seclusion tort is well designed to identify the dignitary injury that results from unwanted advances and other trespasses on personal space. "The gist of the wrong in the intrusion cases is not the intentional infliction of mental distress but rather a blow to human dignity, an assault on human personality,"[49] Edward Bloustein has written. "The woman who is indecently petted suffers from the same indignity as the woman whose birth pangs are overseen."[50] In both cases, the physical intrusions on personal intimacy constitute an assault on the woman's dignity, regardless of whether she suffers economic damages. Courts in intrusion on seclusion cases, like courts in hostile environment harassment cases, have not insisted on proof of serious psychological injury or mental distress.[51] This makes little sense if harassment is conceived as a workplace injury, but it makes perfect sense if harassment is conceived as an indignity and an affront.

The classic examples of intrusion on seclusion involve physical or electronic invasions of traditionally private spaces like the home, and some state courts have held that unwanted advances and other intrusions on someone's psychological solitude cannot qualify.[52] But other courts have concluded that "highly personal questions or demands by a person in authority may be regarded as an intrusion on psychological solitude or integrity and hence an invasion of privacy."[53] The state of Alabama is not generally regarded as being at the vanguard of civil rights, but in a far-reaching decision, the Alabama Supreme Court held in 1983 that a series of intrusive sexual inquiries and repeated sexual demands made by a supervisor in the workplace might constitute an intrusion on seclusion.[54] A plaintiff named Phillips, who worked for a janitorial service in Alabama, suffered a series of invasive personal questions from her boss, who asked her how often she and her husband had sex, what positions they used, and whether she had ever engaged in oral sex. Later he repeatedly asked her to have oral sex with him in his office, covered the windows so that no one else could see in, smacked her bottom as she forced her way out, and eventually fired her. In the Phillips case, the Alabama Supreme Court rejected the argument that an intrusion on seclusion had to be a physical invasion of a private space, like the home. Even in the workplace, the court held, there is an "emotional sanctum" that is entitled to the same expectations of privacy as our private physical space.

The intrusion on seclusion tort has proved to be an effective mechanism for distinguishing serious violations of privacy from unfortunate misunderstandings. The requirement that the indignity must cause "outrage or cause mental suffering, shame, or humiliation to a person of ordinary sensibilities"[55] sets the bar high enough so that relatively trivial indignities are no longer actionable. Juries still face the daunting challenge of trying to abstract away intractable gender differences, and to construct in their minds a gender-blind "reasonable person," but the task is easier than it was in the hostile environment context, because the conduct in question must be "*highly* offensive to a reasonable person" and therefore must be serious enough so that any reasonable person, man or woman, would consider it an unambiguous violation of privacy.

The intrusion on seclusion tort is also well suited to distinguish repeated examples of unwanted touching, which should almost always be illegal, from mere verbal annoyance, which often should not. Alabama courts, for example, have held that repeated sexual touching accompanied by repeated sexual propositions can support a claim for wrongful intrusion.[56] By contrast, they have refused to grant relief for invasions of privacy in a case involving an employer who asked a female employee to "be available," tried to kiss her several times, and later tried to have her fired for resisting his advance (a clear case of quid pro quo harassment). Mere invitations to sexual engagement do not constitute a "wrongful intrusion into one's private activities,"[57] one court held.

Courts have also been more reluctant to impose liability on employers in privacy suits than they are in gender discrimination suits. In Alabama, employers are liable for the intentional torts of their employees only if they know about the conduct, know that it constitutes an illegal invasion of privacy, and fail to take adequate steps to stop it.[58] This provides an incentive for offended employees to complain and for employers to intervene in extreme cases, but it reduces the pressure on companies to monitor private conduct that wouldn't ordinarily be brought to their attention.[59]

Alabama courts have entertained the possibility that cases involving a pattern of hostile and sexually damaging speech may constitute an invasion of privacy.[60] But invasion of privacy law focuses on speech targeted at a particular woman that has the purpose or effect of insulting or humiliating her. This makes it harder for aggrieved third parties to object to dirty jokes, overheard remarks, or pinups in private cubicles. In an invasion of privacy suit, employees could no longer point to the consensual affairs and flirtations and private jokes and e-mail of their colleagues as evidence of a pattern or practice of gender discrimination: the focus would be on the intentional invasion of the privacy of an individual employee, rather than on the social subordination visited on women in general.

It's true, as I mentioned in the Prologue, that unwanted advances may be perceived as forms of gender discrimination as well as invasions of privacy. In the workplace, even relatively trivial forms of sexualization may impede women's progress and make it harder, in subtle ways, for them to function as professionals. (An offense against dignity can also become an offense against equality when it occurs in a professional context.) If the "hostile environment" test were eliminated, lawyers could debate, under either invasion of privacy law or gender discrimination law, whether a particular offense was serious enough to be illegal. But unless Title VII is amended to put the burden of liability on perpetrators rather than on employers, invasion of privacy law gives employers less of an incentive than does gender discrimination law to punish relatively trivial offenses, where the costs of regulation are often greater than the harm that the regulations are designed to prevent.

The tort model would encourage individual responsibility by placing the blame for inappropriate conduct on the perpetrator, rather than the employer, in cases where the conduct took place in private spaces and the employer neither knew about it nor could have been expected to know about it. This would increase the incentives for women to protect their own dignity rather than passively relying on employers to protect it for them. It would empower individual women and men by allowing them to decide whether or not offensive comments interfered with their ability to do the job, rather than using them as involuntary symbols in the gender wars as a whole. It would decrease the economic incentives for plaintiffs to file marginal harassment suits, because individual harassers are less likely to be rich, and therefore less able to pay ruinous damage awards, than deep-pocketed corporations. And given the extensive evidence that men and women perceive the same situations differently, a tort paradigm would increase the possibility of understanding between the sexes by encouraging women or men to complain when they feel offended.

Most important of all, using tort law (rather than the hostile environment test) to regulate conduct that can't be viewed as gender discrimination would make it easier for employers to reconstruct private spaces, inside and outside the workplace, in which employees can express themselves without fear of being monitored and observed. As Erving Goffman has noted, a properly managed workplace will tolerate relaxed "backstage" areas, to give employees a place of refuge from social expectations. "The backstage language consists of reciprocal first-naming, co-operative decision-making, profanity, open sexual remarks, elaborate griping, smoking, rough informal dress, 'sloppy' sitting and standing posture, use of dialect or sub-standard speech, mumbling and shouting, playful aggressivity and 'kidding,' inconsiderateness for the other in minor but potentially symbolic acts, minor physical self-involvements such as humming, whistling, chewing, nibbling, belching and flatulence," Goffman writes.[61] "[B]ackstage conduct is one which allows minor acts which might easily be taken as symbolic of intimacy and disrespect for others present and for the region, while front region conduct is one which disallows such potentially offensive behavior."[62]

Goffman's distinction between the backstage and front region helps us to understand more precisely the relationships between gender discrimination and invasion of privacy. "[T]here seems to be no society in which members of the two sexes, however closely related, do not sustain some appearances before each other,"[63] Goffman notes; he goes on to tell a sunny story about the male workers at a West Coast shipyard who cleaned up their act when they were joined by women during World War II, improving their language, shaving more often, and amiably removing the pornography from their walls and placing it in their toolboxes. Despite its sepia-toned quality—in the wake of the Lewinsky affair, it seems quaint to worry about salacious speech in "mixed company"—the story reminds us that a space where men feel free to let down their hair may be a reasonable woman's gauntlet. When women began to enter the workplace in meaningful numbers, the nature of the backstage had to be renegotiated, in order to create off-duty spaces for men and women to relax together on equal terms.

But creating a genuinely egalitarian backstage space is not so easy. By definition, a backstage area is a place of retreat from social norms—a place where workers are free to be lustful, sloppy, indiscreet, or playful, to form intimate bonds, and to indulge in behavior that would be inappropriate if practiced in more formal areas of the workplace. At a job where men outnumber women in significant numbers, a place where men relax among themselves—making vulgar jokes, talking about women, and engaging in the ordinary rituals of male bonding—may be perceived by women as threatening and discriminatory. When women are in the majority, the backstage area may seem similarly inhospitable to men. Moreover, the spectacle of members of one sex letting down their guard in front of members of the other may itself convey a powerful message of disrespect. Among the Fulani, a cattle-raising society in Upper Volta (now Burkina Faso), in West Africa, men escaped from the elaborate rules of decorum that ordinarily governed community life by visiting a village of former slaves; it was only because the inhabitants of the slave village were considered less than human, and therefore less deserving of respect, that the Fulani felt at liberty to relax in front of them.[64]

There is no simple solution to this problem. If men are prohibited or strongly discouraged from engaging in male bonding in the communal areas of the workplace, they may retreat into less public spaces—moving the backstage area, in effect, into private offices, drinks after works, ball games on the weekend. And the effect of creating a workplace culture where men feel inhibited from making jokes and sexual comments in public places may not be to increase opportunities for women, but to constrict them. In corporations, as older men become wary of forming intimate, mentoring relationships with younger women, for fear of having their attentions misconstrued, they may focus their energies on younger men, with whom they feel more comfortable and less threatened. In universities, male professors go out of their way not to be alone with female students behind closed doors. In workplaces where women vastly outnumber men, similar difficulties may occur.

Striking a proper balance between privacy and transparency in the workplace turns out to be an extremely delicate and complicated task. On the one hand, colleagues ordinarily take care not to learn everything about one another: social interaction is possible only because our whole personality is not put into play, and when colleagues pry too deeply into each other's private affairs there is a risk that one or both of the parties may feel not only that their boundaries have been violated but that their honor has been assaulted. ("An ideal sphere lies around every human being," Georg Simmel writes. "Language very poignantly designates an insult to one's honor as 'coming too close': the radius of this sphere marks, as it were, the distance whose trespassing by another insults one's honor."[65]) On the other hand, without a sphere in which colleagues are free to relax, to form intimate bonds, and even to make mistakes about when an overture of intimacy is likely to be reciprocated, the pressures of social control at work may become overwhelming. This is why I have argued that although backstage behavior by members of one sex can indeed be experienced by members of the other sex as a violation of privacy, the remedy is best negotiated socially rather than legally in situations where the terms and conditions of employment are not clearly affected.

Let me anticipate an objection to my argument that social

disapproval in the workplace is an adequate remedy for those whose privacy is invaded without tangible employment consequences. Self-help for privacy invasions is obviously appropriate in situations where the privacy victim has no fear of reprisals. A former student of mine, for example, was ogled and assaulted with catcalls on the Lower East Side of Manhattan every time she walked to work. In a creative if offbeat burst of inspiration, she bought a disposable camera and every time a man yelled a comment about her appearance, she would surprise him by taking a snapshot. The snapshots themselves are vivid portraits of men experiencing the conflicting emotions of embarrassment at having invaded someone else's privacy and anger at having their own privacy invaded in return. Some of them are narrow-eyed with indignity and resentment; some smile sheepishly; one man hides his face; another poses grandly for his closeup. And one young man, my favorite, blows my student a kiss at the poetic justice of her act, in which the remedy for an invasion of privacy is exactly proportionate to the offense itself.

In the workplace, by contrast, employees will feel less free to confront colleagues or supervisors who have invaded their privacy, for fear of disrupting an ongoing professional relationship. In the interest of maintaining professional comity, employees who feel they have been demeaned or insulted will have an incentive to remain silent, unless the indignity becomes truly unbearable, or unless they decide that the professional relationship is beyond repair. (It is no coincidence that many allegations of sexual harassment surface only after an employee has been fired or decided to quit for reasons that have nothing to do with sex.) But the incentive to overlook minor insults is a characteristic of ongoing social relationships in all small communities, inside and outside the workplace. To the degree that the law now regulates invasions of privacy in the workplace that would be excused or ignored outside of the workplace, it needs to be refined.

If the costs of the current harassment regime were limited to disrupting the careers of a small group of unfortunate employees who have failed to adjust quickly enough to the transformation in social norms that govern interactions between men and women at work, then they would be easy enough for society to bear. All of us should try to behave like ladies and gentlemen in every sphere of our lives, and at this stage in our gender politics, anyone who is reckless enough to treat colleagues, employees, students, and fellow citizens with less than Victorian respect should hardly be surprised by the ignominy that will follow. But the real costs of harassment law are not simply the careers that it disrupts and the workplaces that it fills with confusion and uncertainty. The costs are, more generally, to the boundaries between the public sphere and the private sphere in America.

It is unfortunate, in a liberal society, that sexual expression without tangible employment consequences—overheard jokes, private e-mail, and consensual affairs between colleagues—must be monitored by employers and punished by the state. When speech or conduct changes the conditions of employment, it should be regulated by Federal discrimination law. But in less extreme cases, invasion of privacy law is better suited than the hostile environment test to distinguish speech that is merely offensive from speech that is invasive enough to be illegal. By returning to the text of the Civil Rights Act of 1964, the courts could help to rebuild enclaves of privacy in public places for people to relax, to reveal different sides of themselves in different contexts, to misjudge each other—in short, to be human.

Chapter 4

Privacy in Court

To answer the question that schoolchildren will continue to ask for as long as the Clinton impeachment is studied— "Why did I have to learn so much about Monica Lewinsky's bra hooks?"—you must journey in time back to 1991, when William Kennedy Smith was acquitted of rape. Smith insisted that his accuser had consented to sex but then became angry about something he said, and that only then did she claim that he had sexually assaulted her. The alleged victim testified that Smith had been charming when she met him at a local bar but suddenly turned violent after she accompanied him to the Kennedy compound in Palm Beach. To demonstrate that Smith had a propensity to act that way, the prosecution tried to offer testimony from three other women who said that he had assaulted them in a social setting. But the judge refused to let the jury hear evidence about Smith's sexual past. That refusal outraged many women and built up pressure for legal reform. At the time, Deborah L. Rhode, a feminist legal scholar at Stanford, told a reporter that the Smith trial was "a textbook illustration of what rape victims and sexual-assault victims have long claimed, which is that they're assaulted twice in the process of trying to prove it." Rhode subsequently became the chief counsel for the House Democrats during the Clinton impeachment inquiry.

In Congress, Susan Molinari, then a Republican representative from New York, introduced changes to the federal rules of evidence which would allow juries in sexual-assault and child-molestation cases to consider evidence that the accused had committed similar crimes in the past. (Bob Dole introduced a companion bill in the Senate.) A broad cross-section of judges, legal scholars, and congressional Democrats opposed these revisions, in part because they feared that prosecutors might dredge up earlier offenses for which the defendant had never been charged, let alone convicted. The Federal Judicial Center worried that the new rules would trigger a series of "mini-trials" within the main trial, where the charges and countercharges surrounding each previous bad act would be presented and rebutted. More generally, some opponents of the new rules believed that graphic sexual details were likely to be so inflammatory that a jury hearing them would seek to punish the defendant for what it perceived to be his immoral behavior, whether or not it was convinced of his guilt beyond a reasonable doubt. Moreover, critics of the legislation warned, the proposed definition of "sexual assault"—which included any attempted contact, "without consent, between any part of the defendant's body or an object and the genitals or anus of another person"— was dangerously broad. In their view, Molinari's zeal to "protect the public from rapists and child molesters," as she put it, had produced language that could apply to any fanny-pinching executive, so that the proposed rules would affect not only rape trials but also garden-variety sexual harassment investigations. When Molinari learned that her rule was being invoked in Paula Jones's sexual harassment suit, she professed to be surprised. "That was not my focus at the time," she told me, with a touch of defensiveness.

At first Molinari's proposals drew little support. Then Bill Clinton was elected president. "Clinton basically assisted me in passing that legislation," Molinari told me. In 1994, when Clinton's crime bill seemed hopelessly stalled in the House, she recalled, the President called her and asked what he could do to win her support. She agreed to endorse the legislation if the President accepted her amendments about admitting the evidence of previous offenses by defendants in sex trials. "He told me that he was shocked that it wasn't part of the bill, and he supported it," Molinari said. When the President signed the crime bill, in September 1994, he described its mission in glowing terms. "We together are taking a big step toward bringing the laws of our land back into line with the values of our people and beginning to restore the line between right and wrong," he intoned in the Rose Garden ceremony.

Perhaps Clinton should have paid more attention to the warnings from the legal establishment. By including Molinari's amendments to the rules of evidence (No. 413 in criminal cases and No. 415 in civil cases), he may also, unwittingly, have taken a giant step toward precipitating his own impeachment. At the end of 1997, lawyers for Paula Jones amended their legal complaint to allege that Clinton had "put his hand on Plaintiff's leg and started sliding it toward the hem of Plaintiff's culottes, apparently attempting to reach Plaintiff's pelvic area." The original complaint had not included this detail, and it was added to help establish a sexual assault within the meaning of the new federal rules, so that Jones's lawyers could try to introduce other incidents from Clinton's past. Eventually, investigators working for Jones turned up seven "Jane Does" who were rumored to have had sexual encounters with the President. Though Jones's harassment claim was widely judged to be weak on the law, her lawyers apparently calculated that if jurors heard enough sleazy sexual details about the President they might be inclined to punish him for his bad character regardless of whether they were convinced that his conduct met the legal definition of harassment.

As it turned out, Judge Susan Webber Wright dismissed the Jones suit in April 1998, after concluding that even if Jones's allegations about Clinton's conduct were true they didn't rise to the level of illegal harassment. By then, however, Jones's lawyers had been able to depose Clinton about his contacts with other women and, using a definition that expanded on the language of Molinari's rule, about whether he had "sexual relations" with one of the Jane Does, Monica Lewinsky. Then, five months later, Kenneth Starr submitted an impeachment report to Congress, alleging that Clinton lied under oath and obstructed justice in the Jones case. By filling his report with X-rated details about Clinton's intimate moments with Lewinsky, Starr appeared to be following the same strategy as Jones's lawyers: shore up a questionable legal case with reams of graphic sexual material. As opponents of the Molinari amendments tried to warn Clinton in 1994, it's hard to have a calm debate about legal guilt or innocence when you're distracted by graphic sexual information. This chapter will discuss the unfairness that is inherent in making broad judgments about character on the basis of previous offenses that have been taken out of context.

It seems counterintuitive, on some level, not to let jurors consider evidence from a person's past that might shed light on his or her propensity to commit a particular crime. All of us make judgments about character every day, assuming that if someone is kind, say, in one situation he's likely to be kind in others. Psychologists call the tendency to generalize about a person's character on the basis of a few good deeds the "halo effect." And they caution that such character judgments can be unreliable. In 1968, a psychology professor named Walter Mischel, who now teaches at Columbia, published an influential book called *Personality and Assessment,* which argues that it is hazardous to generalize from behavior in one situation to behavior in other situations. Someone who is highly aggressive in the workplace, for example, could be exceptionally tender with his family. These days, Mischel advocates a contextual view of character, arguing that people may behave honestly at one time and dishonestly at another, because different situations may implicate different abilities, feelings, and beliefs. "Someone who cheats on an arithmetic test might not cheat on a spelling test," he observes.

Mischel's findings bolster the traditional reluctance of English and American courts to allow prosecutors to present evidence that proves nothing but the defendant's bad character—precisely because such evidence can be so prejudicial. "The inquiry is not rejected because character is irrelevant," Justice Robert H. Jackson wrote in a Supreme Court case from 1948 involving a man who had been arrested for receiving stolen goods twenty-seven years before he was charged with bribing federal officials. "On the contrary, it is said to weigh too much with the jury and to so overpersuade them as to prejudge one with a bad general record and deny him a fair opportunity to defend against a particular charge."[1]

Prurient information weighs more than most. Yet in sex cases, courts have sometimes recognized exceptions to the general prohibition of character evidence. For example, English law at one time allowed prosecutors to introduce evidence of deviant sexual history to prove indecency charges. That is why, when Oscar Wilde was tried for committing indecent acts, a parade of his former lovers was called to the stand. Similarly, some American states at one time permitted evidence of past offenses to demonstrate a defendant's "lustful disposition" or "depraved sexual instinct" in certain sex-crime cases. But when Congress set out to codify federal rules of evidence, in the early 1970s, it did not take that approach. Rule 404(a) says, "Evidence of a person's character or trait of his character is not admissible for the purpose of proving that he acted in conformity therewith on a particular occasion."

Past offenses can be, and often are, admitted to prove things other than bad character, such as motive or intent. Nevertheless, a disparity became apparent in sexual assault trials. The sexual history of the accuser could be examined in order to cast light on the question of consent, yet the past behavior of the accused was generally off-limits. Why was it relevant that she had a habit of wearing short skirts, feminists protested, and not that he had a habit of lunging at women who wore short skirts? And so, in 1978, Congress adopted a rape shield law, which protected an alleged victim from being grilled about her sex life. In 1994 this protection was extended to civil cases in which women charged sexual harassment. At the same time, feminists argued that it was unfair to exclude evidence of a man's sexual misdeeds, because a woman's version of what happened in private might not be believed unless her testimony could be corroborated by women who had suffered similar treatment. Persuaded by that argument, Clinton accepted the Molinari amendments in 1994 and thus completed a reversal of the traditional rules of evidence: now the sexual history of the accuser was more or less off limits, while that of the accused could be mined for damaging incidents.

The new protection for the sexual privacy of plaintiffs was hailed by feminists as a necessary response to the Supreme Court's amorphous definition of sexual harassment, which seemed to put the thoughts and feelings of the plaintiff on trial. As I mentioned earlier, the Supreme Court has stressed that, to be illegal, sexual advances and expression in the workplace have to be "unwelcome." The welcomeness requirement is the test that judges use to distinguish consensual from unconsensual encounters at work: not all sexual advances are unwelcome, the courts have reasoned, and it's impossible for a man to know whether or not his advances are welcome unless he asks. Nevertheless, some feminists questioned the welcomeness requirement, claiming that it shifted the focus of harassment trials from the harasser to the victim. The welcomeness requirement also encouraged defendants to invade the privacy of plaintiffs in an effort to argue that the advances in question were reciprocated. In the Meritor case, for example, Chief Justice Rehnquist held that the appellate court had erred by excluding testimony about Mechelle Vinson's personal fantasies and sexually provocative dress. Evidence of "a complainant's sexually provocative speech or dress," Rehnquist held, is "obviously relevant" in determining "whether he or she found particular sexual advances unwelcome."[2]

When Congress, in 1994, amended the federal rape shield rule to apply to sexual harassment trials, it declared that evidence "offered to prove that any alleged victim engaged in other sexual behavior," including her fantasies, dreams, or "sexual predisposition," is generally inadmissible in court. In the wake of the new rule of evidence, No. 412, judges in sexual harassment cases have been more reluctant to admit evidence of the sexual history of the plaintiffs. In a federal case from 1995, for example, Winona Sanchez sued the New Mexico Department of Health, alleging that her supervisor, Mohammad Zabihi, had made unwanted sexual advances toward her and had created a hostile work environment. The Health Department, arguing that Sanchez, not Zabihi, had been the sexual aggressor, asked Sanchez to supply detailed information about any sexual advances she had received from or made to her coworkers at any job during the previous ten years. "In the last (10) years, have you ever . . . had a close personal, romantic, or sexual relationship, however brief, with any co-worker, or any person with whom you worked at the time, or any person who also worked at your same place of employment?" the interrogatory asked. "If so, for each item above, please identify the person(s) involved, the relevant date(s), the relevant place(s) of employment, the number and/or frequency of any such advance(s), whether such advance(s) were welcome or unwelcome, whether you or the other person(s) involved ever complained in any way regarding any such advance(s), and the length and duration of any such relationship(s)." Citing the new federal rules protecting the privacy of plaintiffs, the judge narrowed the scope of the inquiry to the past three years, and held that Sanchez didn't have to answer any of the questions regarding a coworker whom she eventually married.[3]

This seems exactly right. A woman shouldn't have her marital history opened up for public inspection merely because she alleges she has been harassed by one obnoxious coworker. As the federal advisory committee put it, the new rule of evidence, No. 412, "aims to safeguard the alleged victim against the invasion of privacy, potential embarrassment and sexual stereotyping that is associated with public disclosure of intimate sexual details and the infusion of sexual innuendo into the fact-finding process." But the asymmetry of the new federal rules is hard to justify: why should the sexual history of the plaintiff be shielded from exposure while that of the defendant is fair game? In an effort to justify the different treatment of plaintiffs and defendants, the Clinton Justice Department, writing in defense of the Molinari amendments, announced, "Violent sex crimes are not private acts, and the defendant can claim no legitimate interest in suppressing evidence that he has engaged in such acts when it is relevant to the determination of a later criminal charge." But an unwanted grope, although lamentable, hardly qualifies as the kind of "violent sex crime" that should justify the exposure of all the other unwanted advances that an accused man has ever made in his entire life.

Supporters of the Molinari amendments, stressing the improbability that someone would be falsely accused of rape or child molestation on more than one occasion, pointed to a legal theory known as the doctrine of chances. Rather than generalizing about the bad character of the suspect, the doctrine of chances assumes that other allegations of similar acts by the defendant make it more likely that the current allegation is true. The doctrine of chances emphasizes the unlikelihood that someone will be accused of an unusual crime more than once. In an English case around the turn of the last century, for example, when a defendant claimed that his wife had drowned accidently of an epileptic seizure in her bath, the court admitted evidence that two of his previous wives had also drowned in their baths.[4]

Because it's statistically improbable that all three women had drowned accidentally, the earlier incidents made it more likely that the more recent one wasn't an accident.

There is, in fact, a lively statistical debate about whether rapists and child molesters are more prone to repeat offenses than other criminals. Opponents of the Molinari rules argued that the recidivism rates for sex offenders are equal to, or even lower than, those for people who commit other crimes, such as burglary or drug offenses.[5] In 1989, for example, the Bureau of Justice studied 100,000 offenders over three years and found that the recidivism rates were lower for sex offenders than for most other criminals. (The rates ranged from 31.9 percent for burglars to 24.8 percent for drug offenders, 19.6 percent for robbers, 7.7 percent for rapists, and 2.8 percent for murderers.[6]) Furthermore, some feminists criticized the rules of evidence for embracing an unduly rosy view of rape as a pathological activity carried out by a small, mentally ill group of repeat offenders, rather than something that even otherwise healthy men may commit once in their lives.[7] But whether or not the doctrine of chances might justify making a special exception to the rules of evidence in cases involving truly depraved crimes, like rape or child molestation, it is hard to justify as a predictive device in sexual harassment cases, where the distinction between conduct that is boorish and conduct that is illegal may not be obvious until after the case has gone to the jury.

The most prescient objection to the Molinari amendments came from Miguel Mendez, a law professor at Stanford. In comments submitted to the federal judicial conference charged with reviewing the new rules, Mendez cautioned that the halo effect has a corollary, which he called the "devil's-horn effect." If jurors tend to expand a few bits of favorable information into a unified theory of someone's good character, they are even more likely to generalize from past crimes or offensive acts that someone is a bad person and to overlook any exculpatory information.

According to one of the best-known studies of jury verdicts, *The American Jury*, published in 1966 by Harry Kalven Jr. and Hans Zeisel, this effect is particular dramatic in cases involving sexual behavior—which are precisely the cases affected by the Molinari rules. Kalven and Zeisel studied a series of trials in which defendants had engaged in sexual behavior that fell short of the legal definition of a particular crime. "In each of the cases," they concluded, "the jury is so outraged by the defendant's conduct that it overrides distinctions of the law and finds him guilty as charged."

William Miller, the author of *The Anatomy of Disgust*, gives another explanation for the devil's-horn effect, citing the phenomenon of the synecdoche, in which the part comes to stand for the whole. Bad thoughts tend to drive out good thoughts, Miller suggests, and sexual thoughts tend to drive out all other thoughts entirely. The achievement of the Starr report, Miller argues, was to reduce Clinton to an image of nothing more than his grossest impulses. "Who can look at Clinton? All you think about is sexual organs," Miller says. "He is canceled as a moral human being, as someone who needs to be deferred to. He becomes gross matter. There was always this hint about him, overeating, too much hair on his head, something about his fleshiness that made him look almost like Vice in a medieval allegory play."

Miller's study of disgust also suggests that jurors and citizens are more likely to remember kinky sex acts than conventional ones, because unusual acts defy euphemism. "There are certain things that there are no decorous ways of talking about," Miller says. "That's what sank Clarence Thomas. A pubic hair is the proper term. I think although we can turn a blow job into 'oral sex,' and make it go away, once a stain is there the very word stain can't be euphemized. It's wet spots; it's *stains*. It's already irrevocably low and unsalvageable by euphemism." Perhaps for this reason, revelations of extramarital affairs by Republican Representatives Dan Burton, Helen Chenoweth, and Henry Hyde didn't have the same resonance in the public mind. It is far easier to forget a sentence like "I fathered a child" than a sentence like "She also showed him an e-mail describing the effect of chewing Altoid mints before performing oral sex."

Starr insists that he was upholding the rule of law when he decided to flood the country with lurid private information. But did he also preempt any chance for a reasoned public debate about the President's behavior? Although Clinton and Thomas may have been reduced, in the public mind, to a vision of their crudest impulses, the introduction of prurient material into the United States Senate did not distort the public's judgment in precisely the way that the devil's-horn effect predicted. American citizens proved to be more sophisticated than their elected representatives. Rather than succumbing immediately to the urge to punish Clinton as a bad man, a majority of the public drew nuanced distinctions between Clinton's private conduct and his public achievements, resisting the effort to impeach him for misdeeds that, in their view, did not rise to the level of high crimes or misdemeanors. Thomas, similarly, was initially given the benefit of the doubt: although a majority of the public believed that he was telling the truth at the moment of his confirmation, a year later, public opinion had shifted to favor Anita Hill.

In the Senate, President Clinton was acquitted by an almost

entirely partisan vote of 51–49, and Justice Thomas was confirmed by a similarly partisan vote of 52–48. This polarization is consistent with recent studies of social norms. When debate over the objective rightness or wrongness of highly personal conduct is difficult, these studies suggest, people divide themselves into political camps that seem more homogeneous than they really are, in order to signal their broader ideological allegiances.[8] "I don't think there is a standard or sane response to sexual material of this kind," says the political philosopher Thomas Nagel. "Both the material itself and everybody's response to it are highly personal. So it's simply inappropriate to try to introduce those responses into a public forum where agreement is the point."

The polarizing effects of prurience illustrate the central importance of privacy in a pluralistic democracy. In this diverse and truculent country, there are fierce and irreconcilable differences of opinion about the moral, political, and cultural battles that culminated in the Clinton impeachment. In an ideologically divided age, privacy allows citizens to interact on civil terms without confronting areas of fundamental disagreement. When the privacy of public officials is respected, citizens can think each other hopelessly mistaken about questions of sexual conduct, religious obligation, and marital fidelity, yet live together without violent and unproductive clashes that will only exacerbate partisan divisions and make them appear even wider than they already are. When discussions of private conduct and belief are allowed to invade and occupy precious public space, public deliberation may break down entirely, as citizens retreat angrily into ideological camps.

In their essay on privacy, Warren and Brandeis worried that the widespread circulation of prurient information would corrupt and lower public morals. "Each crop of unseemly gossip, thus harvested, becomes the seed of more, and, in direct proportion to its circulation, results in a lowering of social standards and of morality," they wrote. "Even gossip apparently harmless, when widely and persistently circulated, is potent for evil. It both belittles and perverts. It belittles by inverting the relative importance of things, thus dwarfing the thoughts and aspirations of a people." Other liberals of the Progressive era lamented the yellow journalism and realist novels of the 1890s, which published details of adultery and divorce trials and other domestic scandals that would have been unthinkable to discuss in print even a decade earlier. When the details of private affairs are published in the national press, warned E. L. Godkin, the editor of *The Nation,* "a petty scandal swells to the dimensions of a public calamity."[9]

The concern that a sea of prurience will lead to a "lowering of social standards and of morality" seems antiquated today. This is, for better or for worse, a madly sexualized age, in which movies, television, and the Internet ensure that every aspect of popular culture is saturated with images and conversations about sex. The notion that political sex scandals shape public morality more than they reflect it is implausible. It seems more likely that the Clinton and Thomas scandals freed politicians, judges, and journalists to write candidly about sexual practices that most Americans engage in and talk about without embarrassment. "A good effect that probably will be lasting is the encouragement of franker public discussion of sex," Judge Richard Posner writes of the Clinton scandal. "Among the more absurd assertions made in the public debate over the crisis are the Right's charge that the revelation of Clinton's affair with Lewinsky has weakened parental control over the sexual behavior of their children and the Left's charge that the Starr Report is voyeuristic and pornographic. People who say these things (and mean them, as perhaps few do) don't understand the family and sexual culture of late twentieth-century America."[10]

Brandeis and Warren's more relevant objection is that the prurient information is so luridly interesting that, when widely publicized, it crowds out all other topics of public discussion, making it difficult to think or talk about anything else. "When personal gossip attains the dignity of print, and crowds the space available for matters of real interest to the community, what wonder that the ignorant and thoughtless mistake its relative importance," they wrote. "Easy of comprehension, appealing to that weak side of human nature which is never wholly cast down by the misfortunes and frailties of our neighbors, no one can be surprised that it usurps the place of interest in brains capable of other things. Triviality destroys at once robustness of thought and delicacy of feeling. No enthusiasm can flourish, no generous impulse can survive, under its blighting influence." Because the space available for public discussion is limited, privacy serves as a shield against what social scientists have called the "availability heuristic," which holds that, in order to engage in rational thought and discussion, we need information screens to prevent our minds from becoming overwhelmed by the mass of information that competes for our attention.[11] "It's not inconsistent for the public to say 'I wish it would go away' and 'I wish it had never come up' but also to be unable to resist reading about it and allowing it to crowd everything else out," says Thomas Nagel.

Since national sex scandals are relatively infrequent, the phenomenon of lurid material crowding out all other topics of public discussion won't bring the government to a halt. The more lasting legacy of the devil's-horn effect is that knowing every-thing about someone's private life inevitably distracts us from making reliable judgments about his or her character and public achievements. E. L. Godkin, for example, deplored the loss of proportion that ensued when gossip circulated outside the con-text of a community of people who were acquainted with the parties involved and could judge their behavior. Tabloid gossip, he sniffed, would commit "a fraud on the public" by giving the opinions and wishes of obscure and unimportant people "an amount of respect . . . to which they are not entitled," at the same time depriving "men of real value of their proper place in the public estimation."[12]

Although Godkin's concerns seem unfashionably aristocratic today, Washington is full of public servants whose reputations have been unfairly stained by the exposure of relatively small transgressions. Clarence Thomas is a vivid illustration. After nearly a decade on the Supreme Court, he has developed into a provocative, fiercely independent, and interestingly radical jus-tice, willing to rethink entire areas of constitutional law from the ground up. But in the public mind, he remains nothing more than a dirty joke. In *Resurrection*, John Danforth's oddly pruri-ent memoir of the confirmation ordeal, Thomas reflected elo-quently about the unfairness of being reduced to the sum of his most embarrassing impulses. "It showed me just how vulnerable I am. . . . It showed me how vulnerable any individual is," Thomas told his friend Danforth. "It also showed me some-thing that I will never lose, and that is that 2 percent or 1 per-cent or 0.2 percent can always be used to destroy a human being when there are no barriers, when there is no perspective and no context."[13]

Viewed through the prism of the Clinton impeachment, the

Hill-Thomas hearings look far different than they did in 1991. Even those who believe that Anita Hill was telling some version of the truth when she alleged that Thomas had asked her out on dates and discussed pornographic movies must feel qualms about having condemned Thomas's denials as categorically as many of us did at the time.[14] If Clinton's lies about his encounters with Monica Lewinsky were justified or mitigated by the claim that he shouldn't have been asked to discuss such private matters in the first place, shouldn't Thomas be excused for resorting to a similar form of self-exculpation? Clinton himself came to understand the moral complexity of Thomas's response to Hill when he tried, during his own grand jury testimony, to explain the apparently irreconcilable contradiction between Monica Lewinsky's testimony and his own in recalling the sexual details of their encounter. "This reminds me, to some extent, of the hearings when Clarence Thomas and Anita Hill were both testifying under oath. Now, in some rational way, they could not have both been telling the truth, since they had directly different accounts of a shared set of facts," Clinton said. "When I heard both of them testify, what I believed after it was over, I believed that they both thought they were telling the truth. This is—you're dealing with, in some ways, the most mysterious area of human life."

Critics of Clinton and Thomas insist that the relevant question was perjury, not sexual harassment, and that it was the unequivocal denials, not the sexual misconduct, that justified the public exposure. But people often lie when interrogated about sex, and American law used to be more sensitive to human frailty than it is now. For most of American history, courts didn't put people under oath in situations where they might be tempted to perjure themselves, and judges also distinguished among different kinds of lies, examining the liar's state of mind, the seriousness of the lie, and its effects on other people. There is no legal entitlement to lie about sex, nor should there be. But if it was inappropriate to ask Clinton and Thomas about their private conduct in the first place, self-exculpatory lies designed to protect privacy shouldn't provide a moral justification for the evisceration of privacy.

It seems particularly unfair that Thomas's privacy should be turned inside out by allegations of misconduct that can't clearly be identified as illegal harassment, even ten years after the fact. This is a consequence of the ambiguity of harassment law itself. At the time, Hill stressed that although she believed she had been the victim of legally actionable harassment, she recognized that others might disagree because the law was in flux. "I would suggest that saying that it is sexual harassment and raising a legal claim are two different things," she told Senator Arlen Specter. "What I was trying to do when I provided information to you was not say to you, 'I am claiming that this man sexually harassed me.' What I was saying and what I state now is that this conduct took place." But even in light of the Supreme Court's subsequent refinements of the law, it's still not clear whether or not Hill was sexually harassed in the legal sense. She claimed that during the time that she worked for Thomas at the Department of Education and the Equal Employment Opportunity Commission, between 1981 and 1983, Thomas asked her out on dates between five and ten times, discussed pornographic movies, told a vulgar joke, and boasted about his sexual prowess, without ever explicitly asking her to have sex with him.

Was this conduct "severe or pervasive enough to create an objective hostile or abusive work environment—an environment that a reasonable person would find hostile or abusive," to use the Supreme Court's current test? None of the Court's suggestions for judging hostility or abusiveness—the "frequency and severity of the conduct; whether it was physically threatening or humiliating as distinguished from a mere offensive utterance" and "whether it unreasonably interferes with an employee's work performance"—are particularly helpful in providing an answer. Are five or six requests for a date during a two-year period frequent or severe? Was the vulgar joke about the Coke can abusive and humiliating or merely offensive? Hill says that she was hospitalized for five days for acute stomach pains, which she attributed to stress on the job, and that she feared that Thomas would retaliate against her for rebuffing his advances. But in fact, Thomas told her he was pleased with her work and supported her effort to begin an academic career. Hill, who was more ambitious than she acknowledged, accepted Thomas's support and continued to rely on him as a mentor and an advocate. Most important of all, Hill never invoked the grievance procedure at the EEOC or filed a formal complaint, a choice that she called "maybe . . . poor judgment, but it wasn't dishonest and it wasn't a completely unreasonable choice that I made given all the circumstances." In light of the Supreme Court's most recent pronouncements, Hill's failure to complain might by itself doom any subsequent harassment suit and protect her employer from legal liability.

Held in the Senate, the Thomas hearings were conducted without the protections of the rules of evidence that apply in an ordinary harassment trial, and yet it is surprising how precisely the unconstrained hearings anticipated the way actual harassment trials would come to be conducted under the new rules of evidence. In a crude effort to discredit Hill, Thomas's supporters tried to prove that she was a sexual fantasist: on the basis of a single encounter with Hill, John Doggett memorably alleged in an affidavit that, in his opinion, "Ms. Hill's fantasies about sexual interest in her were an indication of the fact that she was having a problem being rejected by men she was attracted to."[15] In the same spirit, several former law students at Oral Roberts University were moved to swear that Hill had put pubic hairs in their exam booklets. Even Bush administration officials, according to Thomas's sponsor, Senator John Danforth, found this absurd, and Danforth confessed in his memoir that he now regrets the unhinged ruthlessness with which he tried to discredit Hill, without showing any concern about fairness or evidence.

Thomas's privacy was also shattered, with similarly brutalizing consequences. In order to corroborate Hill's claim, her supporters tried to present evidence of other previous sexual misconduct for which Thomas had never been charged. They offered an affidavit by Angela Wright, a former EEOC employee who claimed that Thomas had a peremptory way of inviting her on dates: "You know you need to be dating me—I think I'm going to date you." Although this contradicted Thomas's claim that he had never asked out anyone on his staff, and it echoed language that Hill said Thomas had used with her, Wright herself acknowledged that this might not be considered legally actionable harassment. Most disgracefully of all, Senate investigators and journalists sifted through Thomas's private magazine collection and video rental records, and the titles of movies he had allegedly rented were later broadcast on national television.

Evidence that Thomas had a habit of watching pornographic movies, and talking graphically about them among friends, was not exactly irrelevant to corroborating Hill's charge that Thomas had a habit of discussing pornographic movies in the workplace. The question is whether the charges were serious enough to justify the assault on Thomas's privacy that resulted from the effort to prove them. The sin of having discussed pornography with an employee in his office was hardly so grave that it merited the exposure of Thomas's most intimate thoughts and fantasies to the entire world.

As I suggested earlier, Thomas seems to have been guilty not of gender discrimination—his record of promoting women in general was beyond reproach, and he helped Hill repeatedly after she rejected his advances—but of invading Hill's privacy. Yet if Hill had sued Thomas for intrusion on seclusion, her suit could have been summarily dismissed without an intrusive discovery process. Even assuming Hill's allegations to be true, it is not clear whether Thomas was engaging in a clumsy but well-intentioned courtship ritual that she never made clear offended her. If we take the dimmest view of Thomas's conduct, we might surmise that he singled Hill out for special attention precisely because of her prudish and reserved manner: he knew that she would be especially embarrassed by pornographic talk, and he took a perverse pleasure in discomfiting her. Even so, courts have held that vulgar jokes and unwanted requests for dates, unaccompanied by unwanted touching, would not be offensive enough to "outrage or cause mental suffering, shame, or humiliation to a person of ordinary sensibilities."[16] The talk about kinky pornographic movies was the most obvious violation of Hill's privacy, but dirty talk has been actionable under the privacy torts only when it is part of intrusive questions about the employee's own private life. Assuming Hill's allegations to be true, in short, if the Senate had treated Hill's allegations like an ordinary suit for invasion of privacy, it could have dismissed the charges, and avoided the need to violate Thomas's and Hill's privacy in the process.

If the Thomas hearings provided a preview of the invasions of privacy that are inherent in harassment law, the Jones case provided a full-scale confirmation. In dismissing Jones's case as without legal merit, Judge Susan Webber Wright assumed that Clinton had indeed exposed himself to Jones in a hotel room, asked her to "kiss it," and then retreated when he was rebuffed. Even if Jones's allegations were viewed as favorably as possible, however, Judge Wright found that Jones still hadn't made out a plausible case of quid pro quo harassment, because she suffered no tangible job detriments for having rebuffed Clinton's advance. And Judge Wright found Jones's claim that she suffered from hostile environment harassment to be similarly questionable. It's rare for a single sexual advance to be punished as sexual harassment. Because men can't know whether or not an advance is unwanted unless they ask, courts have tended to give employers one bite at the apple. The EEOC's most recent Policy Guidance on Sexual Harassment says that "unless the conduct is quite severe, a single incident or isolated incidents of offensive sexual conduct or remarks generally do not create an abusive environment." Because Clinton's self-exposure was not "frequent, severe, or physically threatening," Judge Wright concluded, it couldn't be considered pervasive enough to have altered the conditions of Jones's employment.

Given the amorphousness of the hostile environment standard, a jury, or an appellate court, might have reached a different conclusion than Judge Wright about whether a reasonable person would have found Clinton's conduct hostile or offensive. In their appellate brief, Jones's lawyers noted a recent decision by the federal appeals court in Little Rock in which the judges observed: "We cannot say that a supervisor who pats a female employee on the back, brushes up against her, and tells her she smells good does not constitute sexual harassment as a matter of law."[17]

The Jones case showed that even a legally questionable allegation of harassment can result in dramatic invasions of the privacy of innocent people during the period before the case is ultimately dismissed. In October 1997, as I discussed earlier, Jones's lawyers amended her complaint to allege that Clinton was a "sexual predator." He had discriminated against Jones "because of her sex," they argued, by granting employment benefits to other women who succumbed to his advances, while denying similar benefits to Jones because she rebuffed him. To support their theory, they asked Clinton to name all women other than his wife with whom he had "proposed having" sexual relations during the nearly twenty years that he was attorney general of Arkansas, governor, and president. The audacity of the claim was striking: as Clinton's lawyers pointed out, courts have held repeatedly that it isn't illegal for an employer to favor his paramour over other employees, and they asked Judge Wright to limit discovery in the case "to purported incidents, if any, of non-consensual conduct."

But in a crucial response drafted by the two anonymous Jones advisors who would later introduce Linda Tripp to Kenneth Starr, Jones's lawyers listed a number of reasons that they should be able to pry into Clinton's consensual affairs. First, they noted, the new federal rule of evidence, No. 415, makes similar acts of sexual assault admissible to show that the defendant had a propensity for abusing women. This seemed like a stretch. Rule 415 only authorizes the admission of previous acts of sexual assault, while Jones argued that she should be able to rummage through Clinton's consensual affairs. But Jones's lawyers insisted that "there is no practical means for this Court, in advance, to limit discovery to non-consensual sexual behavior because only after discovery can the existence of consent be determined."[18] With a MacKinnonite flourish, they added that sexual consent "is extremely difficult for anyone to define" and that Clinton "may have convinced himself that a woman consented when in fact she did not."[19]

Because she had already ruled that Jones was not the victim of attempted sexual assault, Judge Wright could have rejected Jones's request to rummage through Clinton's consensual sex life. But a month later, she granted part of Jones's request, ruling that Clinton had to identify any state or federal employees with whom he had sexual relations, or proposed to have sexual relations, during the five years before or after May 8, 1991, the date that Paula Jones was allegedly traumatized in the Excelsior hotel. "The Court cannot say that such information is not reasonably calculated to lead to the discovery of admissible evidence,"[20] Judge Wright observed, accurately citing the permissive discovery standard. At the President's deposition in January, when Jones's lawyers presented Clinton with a definition of "sexual relations" that closely followed the language of Rule 415, Clinton's lawyer, Robert Bennett, objected that part of the definition could be construed to encompass nonconsensual contacts between two people. "[I]f the president patted me and said I had to lose ten pounds of my bottom, you could be arguing that I had sexual relations with him," Bennett argued. Agreeing with Bennett, Judge Wright stressed that she was interested in "contact that is consensual" and she restricted the definition to include contact with the private parts "of any person with an intent to arouse or gratify the sexual desire of any person." In restricting the definition, she observed, "The Court has ruled that consenting contact is relevant in this case."[21]

The chaos that followed from that ruling showed the folly of allowing courts to scrutinize purely consensual affairs in the workplace. As critics of the Molinari rules had warned, the discovery process triggered a series of mini-trials as the most minute details of Clinton's interactions with each of the Jane Does were scrutinized to determine whether or not the encounters were consensual, and whether or not they had resulted in job benefits or detriments. Given the ambiguities of sex in the workplace, both of those questions proved, in each case, hard to answer.

After Judge Wright excluded evidence of Clinton's affair with Lewinsky on the grounds that it wasn't essential to the core issue in the case—namely, whether Jones herself had suffered from illegal sexual harassment—Jones appealed the ruling, arguing that Clinton's affair with Lewinsky couldn't be considered consensual because of the power imbalance between the two parties, and therefore it had to be admitted under the Molinari rules of evidence. "If any aspect of Defendant Clinton's sexual conduct with respect to Ms. Lewinsky was not consensual, then it is admissible under Fed. R. Evid. 415," Jones's lawyers wrote. "Sexual advances by the President to a 21-year-old intern who is working in the White House are inherently coercive." Under this analysis, any affair between a powerful man and a younger woman is a sexual assault, even if both of the parties concerned believe that it is consensual.

The Jones lawyers justified their questions about Monica Lewinsky by insisting that her ordeal would help them prove their case that Clinton conditioned job benefits on sexual favors. But once again, the facts failed to support their theory. Rather than providing an example of how an employee who succumbed to Clinton's advances was rewarded with a government job, Monica Lewinsky's experience showed how an employee who initiated an affair with the President was transferred out of government. But this wasn't obvious until after Clinton's affair with Lewinsky had been exposed, and the damage to Lewinsky's privacy was irreparable.

Before she became a household name, Lewinsky filed a motion to quash her subpoena, in an effort to avoid having to testify in the Jones case. The motion itself is a sobering testament to the poverty of existing legal protections for the privacy of intimate secrets. "The discovery process may seriously implicate privacy interests," Lewinsky's lawyer Frank Carter wrote. "Surely if plaintiff has the resources and the time, she could depose each and every woman who ever worked at the White House during the administration of Defendant Clinton, all in an attempt to see if there was any shred of rumor or innuendo which she might uncover. This would have the impact, as this deposition is having on Jane Doe #6, of invading the privacy of each woman and causing each to expend money for counsel fees and costs."

Lewinsky was especially upset about the private documents demanded by Jones's subpoena. "They want, among other things, 'every calendar, address book, journals, diaries, notes, letters, etc.,' " Frank Carter objected on January 20, 1998.[22] Arguing that Jones was trying to "unreasonably invade" Lewinsky's privacy and "disrupt her private life," Carter asked Judge Wright to quash the subpoena. Before Judge Wright had a chance to rule on Lewinsky's motion, however, Kenneth Starr intervened.

If Paula Jones had sued Clinton for invasion of privacy, rather than sexual harassment, the unnecessary violations of Lewinsky's privacy would have been averted. Like Anita Hill's hypothetical suit for intrusion on seclusion, Jones's suit could have been summarily dismissed before discovery began, although the invasion that Jones suffered was arguably greater than that of Anita Hill. Having someone unexpectedly expose himself in a hotel room might be considered a more dramatic violation of a woman's personal boundaries than repeated requests for dates. In one recent case, for example, an Alabama court said that a reasonable jury might have found that the employee suffered an intrusion on her seclusion when her boss invited her to a hotel purportedly to review business papers relating to a theft at a local outlet. After telling her that her job was not in jeopardy, he wrapped his arms around her waist; he then shut and locked the door, pulled her toward the bed, and when she headed for the door a third time, tried to hug and kiss her. After the incident he made daily comments about her appearance; and when she tried to leave his office, he grabbed her and told her to sit down.[23]

Clinton behaved more grossly if he exposed himself as Jones alleged; and even if we assume that he touched Jones only once, and retreated after he was rebuffed, many women would be genuinely and understandably distressed and traumatized by such a crude advance. But a gross exposure isn't ordinarily legally actionable as an invasion of privacy. Even if a judge had allowed discovery in a hypothetical invasion of privacy suit against Clinton, the discovery process would have focused on Clinton's treatment of Jones, rather than his treatment of all the women to whom he had made sexual advances, and the privacy of innocent third parties would be have been preserved.

If Hill and Jones had sued Thomas and Clinton for invasion

of privacy, judges sensitive to privacy concerns would be faced with hard questions about whether to exclude certain evidence as more prejudicial than probative. Evidence of Thomas's video rental habits, or Clinton's proclivity for making passes at women in the workplace, would not be entirely irrelevant in a hypothetical invasion of privacy suit, but an alert judge could exclude the evidence, balancing its probative value against its potential for invading privacy. And because the invasion of privacy tort, unlike the law of sexual harassment, focuses on a particular encounter rather than a pattern of behavior, a sensitive judge would have more discretion to exclude invasive evidence without making it impossible for the plaintiffs to put on their case.

Congress could enhance this judicial discretion by amending the federal rules of evidence so that accused harassers enjoy the same protections against being judged on the basis of their past sexual misconduct as alleged victims. (At the very least, the definition of "sexual assault" in the new rules of evidence should be refined so that it does not apply to garden-variety harassment cases.) Congress could also create a sexual privilege that protects parties and witnesses in harassment cases from having to discuss their consensual sexual history. Like the attorney-client privilege or the psychotherapist-patient privilege, the consensual sexual privilege might be overcome in extraordinary cases, involving other serious crimes or imminent harm. A privilege along these lines, would, of course, make some sexual harassment allegations harder to prove. But throughout American history, legislators have balanced privacy claims against the seriousness of the offense in question, and have concluded that it is worth tying the hands of prosecutors in categories of cases in which uncovering guilty activity requires the exposure of a great deal of innocent activity.

Indeed, sexual harassment suits today share at least some of the characteristics of the heresy and blasphemy prosecutions that the framers of the Constitution were willing to under-enforce in the interest of protecting privacy. In England, blasphemy was one of the four branches of criminal libel—the others were obscenity, sedition, and defamation—all of which covered dignitary offenses and were designed to ensure that speech didn't violate norms of respect and propriety.[24] The distinction between legal and illegal speech could turn not only on the substance of the speech but on the outrageousness of its manner of expression, which meant that it was often impossible to know before trial what, precisely, juries would condemn. In the twentieth century, similarly, harassment law was designed to protect the reputation of women in general, and it achieves its purpose by forbidding the verbal denigration of women in some terms but not others. Because harassment suits often turn on subjective reactions that are hard to verify, allowing plaintiffs in harassment suits to subpoena the private papers and diaries of coworkers potentially exposes the innocent but embarrassing secrets of a great many citizens to public view—precisely the invasion of privacy that Lord Camden feared in libel cases. For this reason, a Congress sensitive to the history of the Fourth Amendment might conclude that constitutional values support the creation of a sexual privilege.

I've argued in this chapter that when prurient information is introduced in court, it can distract jurors and the public at large, leading them to judge the accuser and the accused on the basis of embarrassing past acts rather than their guilt or innocence of the charges in question. Let me now consider a potential objection to my argument. By focusing on unusual national spectacles such as the Clinton and Thomas scandals, perhaps I have chosen cases that illustrate the persistence of privacy in America, rather than its erosion. In addition to being appropriately subjected to forms of surveillance that ordinary citizens seldom experience, the objection might go, Clinton and Thomas are figures in whom and through whom the relationship between norms and law was publicly elaborated. The public, alternately repelled and fascinated by the spectacles, may have taken prurient pleasure in witnessing the violation of its own norms, but was ultimately able to make subtle and nuanced distinctions between the entertaining melodrama and the political outcome. Public trials, as Émile Durkheim recognized, have a strong ritual element, and the process of identifying and punishing exemplary violations of moral and social rules is the process by which we identify what, precisely, those rules are. By insisting that Clinton be acquitted and Thomas confirmed, perhaps the public established clear limits on what kinds of violations of privacy it would tolerate, and, in the process, it reasserted the boundaries between the public and private sphere.

In hindsight, however, the social meaning of the Clinton and Thomas ordeals is not so easy to identify. Although Thomas was confirmed, his reputation in the eyes of the general public was seriously damaged, while Clinton's popularity after his acquittal by the Senate remained high. The very different public images of Clinton and Thomas illuminate, I think, another aspect of the phenomenon of the synecdoche: those with unlimited access to the public's attention have a greater chance of being judged in context than those who do not. Because Justice Thomas is compelled by the conventions of his office to remain fairly secluded, he has had less opportunity than President Clinton to reveal other aspects of his identity and to correct the damage that public humiliation did to his reputation. Similarly, it is hard to think of President Ford without recalling his tripping at the airport, because Ford's presidency was so short and produced few other memorable images. But it is possible to think of President George Bush without recalling his becoming ill at the state dinner in Japan, because Bush was able to define himself in other ways (Gulf War hero! no new taxes!) that compete in the public memory with the spectacle of his indisposition. Private citizens, I will argue in the next chapter, are more like Justice Thomas and President Ford than like Presidents Clinton and Bush: we have fewer opportunities to present ourselves publicly in all of our complexity. Therefore, as more of our private lives are recorded in cyberspace, the risk that we will be unfairly defined by isolated pieces of information that have been taken out of context has increased dramatically.

Privacy is a duty as well as a right, some have suggested, and in order to have the freedom to behave unconventionally in private, one should take care not to flaunt unconventional behavior in public.[25] By this reasoning, it might be argued that when public officials or private citizens behave recklessly, the recklessness itself creates a public issue out of what ordinarily should be considered private conduct.[26] But this confuses an individual's personal interest in privacy with the public's interest in not being distracted by prurience. Congress should restore the frayed boundaries between sex and politics, in other words, not for the president's sake, but for ours.

Chapter 5

Privacy in Cyberspace

In 1998, the dean of the Harvard Divinity School was forced to resign after he downloaded pornography on his home computer. The dean had asked a Harvard technician to bring a computer with more memory to his official residence, and in the course of transferring files from the old computer to the new one, the technician noticed that the dean had saved thousands of pornographic pictures. Although none of the pictures appeared to involve minors, the technician told his supervisor, and word eventually filtered up to the president of Harvard, who asked the dean to step down.

When the story broke several months later, Harvard justified its decision to force out the dean by noting that the rules of the Divinity School prohibit the personal use of university computers in ways that clash with the school's educational mission. More specifically, the rules forbid the "introduction" of material that is "inappropriate, obscene, bigoted or abusive" into the educational environment.[1] But the dean was at home, not at school, and the fact that the university owned his computer and his house hardly settles the question of whether his personal browsing was anyone else's business. If the dean had made long-distances calls to his mistress on a Harvard-owned telephone, most people would consider university monitoring of his conversations an unsettling invasion of privacy. And although private universities aren't formally required to respect the First Amendment, it would be hard to imagine Harvard forbidding members of the Faculty of Arts and Sciences from downloading "inappropriate" material in their own homes, even on university-owned computers. If the dean had been an aficionado of Snoop Doggy Dogg, no one would have suggested he should be fired for downloading gangsta rap.

The best defense of Harvard's decision is one that university officials never made openly: divinity schools, perhaps, are different from law schools. The dean was an ordained minister in the Lutheran Church of America, which has issued a pastoral statement condemning the consumption of pornography as a misuse of sexuality, a form of self-degradation that violates the more general proscription of sexual relations outside of marriage. The Harvard Divinity School, however, is not a seminary, and it's not clear that secular universities should be in the business of disciplining faculty for breaching their sectarian commitments.

Moreover, the rise of cyberspace has blurred the distinction between home and office, as faculty increasingly use university-owned computers and networks at home as well as at work. I'm at home, for example, as I type these words, but the computer on which I'm typing is owned by the law school where I teach, as is the network that supplies my e-mail and Internet access. I would be appalled if anyone suggested that the provision of these research tools gave my law school the right to monitor everything I read and write. Twenty-four-hour surveillance may be appropriate in a company town, but it threatens the central purpose of a university: to promote freedom of thought. And even in a company town, there should be some zones of privacy from corporate scrutiny. In 1880, for example, George Pullman founded the town of Pullman, Illinois, where he owned all the public and private spaces, from the hotel, church, and fire department to the gasworks and waterworks. Pullman's inspectors enforced rigid codes of conduct by fining citizens who misbehaved, and democratic elections were not permitted. In 1898, however, the Illinois Supreme Court ordered Pullman Inc. to divest itself of most of its property in Pullman, Illinois. In America, the court held, you can own a company, but not a town.[2]

The real lesson of the Divinity School scandal wasn't legal but technological: the dean's downfall reminds us how much of our reading habits on the Internet are exposed to public view. Not long after the scandal, a former computer technician at Harvard Divinity School wrote an article for *Salon,* the on-line magazine, criticizing his former colleagues for snitching on the dean. Writing under a pseudonym, the author acknowledged how much he routinely learned, in the course of doing his job, about the most intimate activities of other members of the university community. Computer caches and Internet history files reveal the pictures that users have downloaded and the sites they have recently visited. Even when the caches and history files are cleared, reading habits can be reconstructed by "cookies" embedded in a user's hard drive.

"In the server room of one of my part-time jobs," the techie confessed, "I noticed that a program called Gatekeeper displayed all the Internet usage in the office as it happened. I sat and watched people send e-mail, buy and sell stocks on E*trade and download pictures of Celine Dion. If I had wanted I could have traced this usage back to the individual user."[3] Although everyone in the company signed the same form consenting to have his or her Internet use monitored, the techie knew that the employees, consciously or unconsciously, expected that they weren't being watched, in the same way that one makes phone calls without worrying about wiretaps. For this reason, the author considered it a point of courtesy to respect the privacy of the employees he helped, much like the Secret Service agents who accompany the president at all times and are trained not to listen to his conversations:

> [T]here is a way to look without looking, to help someone with a Word document without actually reading it, to troubleshoot a system for every conceivable problem and not notice the things on a hard drive. This requires a carefully honed attitude, a way of thinking that what is yours is yours and mine is mine. Not everything you do all day long is my business—and if I happen to see something you would rather I didn't, I extend you the same courtesy I want extended to me.[4]

Computer technicians aren't the only people who have the capacity to monitor our reading habits on the Internet. Before Harvard tightened its access rules in 1995, the school's network was set up so that any user could type a command at his or her console and be given a detailed report on what every other user of the system was reading, downloading, or e-mailing at that particular moment. (One Harvard student threatened suicide and sought psychological counseling when he realized that the student newspaper had detailed records of his pornographic browsing.)[5] On many computer networks, it's not hard for a computer-savvy snoop to type a command like "ps-aux" or "w" at the command prompt that will reveal all the users who are logged on to the system, along with the programs they are running. It's also simple to get a program that will "sniff" the packets of data coming from other computers, or scan the system periodically, in order to collect detailed profiles of what others are reading and thinking, and to whom they are talking.[6]

But network snoops are only the tip of the iceberg. In cyberspace the greatest threat to privacy comes not from nosy employers and neighbors but from the electronic footprints that make it possible to monitor and trace nearly everything we read, write, browse, and buy. Most Web browsers are configured to reveal to every Web site you visit the address of the page you visited most recently and your Internet Protocol address, which may—or may not—identify you as an individual user.[7] This information can be collected and stored to create detailed profiles of your tastes and preferences and reading habits, which are highly valued by private marketers. Some Web sites send e-mail advertisements to everyone who accesses their pages, and the largest Web sites use advertising brokers to track the other Web sites each user visits. These brokers can send cookies to your hard drive that track a host of personal information, from the other sites you visit to keywords from your AltaVista searches. If you then buy something from another site that subscribes to the same advertising broker, the broker can match your name to the cookie and sell the profile to direct marketers, without your knowledge or consent.[8] This accounts for the unsettling experience some Internet users have had of being bombarded with targeted ads after expressing an interest in a particular topic. It's technologically possible, and legally permissible, for a drug company to send an unsolicited ad for a new prostate cancer treatment to someone who surfs a Web site about prostate cancer and then buys a book on the subject.[9] As long as users were confident that their virtual identities weren't being linked to their actual identities, many were happy to accept cookies and targeted ads in exchange for the convenience of being able to navigate the Web more easily. Then DoubleClick Inc., the Internet's largest advertising company, bought Abacus Direct, a database of millions of names, addresses, and information about off-line buying habits compiled from the nation's largest direct-mail catalogues. When DoubleClick announced that it was planning to compile dossiers linking individuals' on-line and off-line purchases and browsing habits with their actual identities, its stock plunged and DoubleClick announced that it would delay its plan. But in the near future, there is a real danger that personally identifiable dossiers of our on-line and off-line activities could be compiled, sold, or subpoenaed without our consent.[10]

Direct marketers are not the only people who have an interest in your consumer profiles and technological footprints. Your insurance company might want to know if you have been browsing books about sexually transmitted diseases; your employers might have an interest in your on-line investments; and the FBI might want to keep an eye on you if you display an unhealthy interest in the oeuvre of Timothy McVeigh. In real space, the sale of consumer profiles to the government has already led to some alarming linkages. Farrell's Ice Cream, for example, sold the names of customers who received free sundaes on their birthdays. A marketing firm bought the list and sold it to the Selective Service System, which promptly sent draft registration warnings to a group of ice-cream lovers who had come of age.[11] In cyberspace, however, far more sophisticated linkages are becoming routine, in a world where more and more of our transactions are conducted on-line. And unlike the European Union, the United States has no comprehensive privacy law prohibiting data collected for one purpose from being sold and used for another.

To help dramatize the new architecture of privacy, Professor Jerry Kang of UCLA Law School asks us to imagine visits to two malls, one in real space, the other in cyberspace. In a real mall, he notes, you might buy an ice-cream cone with cash, flip through magazines in a bookstore, and then buy a silk scarf with a credit card. Strangers may notice you passing, but are unlikely to remember precisely what you were doing and when you were doing it, and human memories quickly fade. The only exception is your purchase of a scarf with a credit card, which generates data that are detailed, traceable to you, and potentially permanent.[12] By contrast, Kang notes, "in cyberspace the exception becomes the norm: Every interaction is like the credit card purchase."[13] Invisible scanners automatically record the stores you visit, the magazines you browse, the pages you linger over, and the time that you linger over them. As with the credit card purchase in real space, all the click-stream data generated in cyberspace are permanently traceable and permanently retrievable.

And Kang's dystopia is only a preview of things to come. Changes in the way books and movies and music are delivered will pose an even more serious threat to what one author has called the "right to read anonymously."[14] The same technologies that are making it possible to download digitally stored books and CDs and movies directly from publishers onto our hard drives are also making it possible for publishers to record and monitor our reading habits with unsettling specificity. Copyright management systems can regulate not only which books you read but how many times you read them, charging different royalties based on whether you copy from the book or forward part of it to a friend. If you download a collection of short stories, your computer can notify the author every time you read a particular story, preventing you from reading it again until you have paid the agreed fee. As newspapers are increasingly delivered on-line, customers can be charged different prices for how much of the newspaper they read, with each downloaded article tagged and recorded. This detailed model of browsing habits can be used to create reader profiles that can, in turn, be sold at a premium.

Even television is now being redesigned in a way that creates records of our viewing habits. A new electronic device known as a personal video recorder makes it possible to store up to thirty hours of television programs and record shows at the same time. Some network executives fear that within ten years, the device will change the way television programs are delivered, making it easy for viewers to skip commercials and thus ensuring that most programs are sold by studios or by their creators to viewers on a pay-per-view basis. One of the current models, TIVO, asks viewers about their age, income, and interests, establishing viewer profiles that it then uses to record future shows accordingly to previously established preferences.[15]

The monitoring of reading, listening, and viewing habits is something that we are reluctant to tolerate outside of cyberspace. But why? The usual answer is that our freedom of thought is violated when our reading habits are monitored, but this isn't entirely convincing. I'm free to think whatever I like even if the state or the phone company knows what I read. Instead, people are reluctant to have their reading and viewing habits exposed because we correctly fear that when isolated bits of personal information are confused with genuine knowledge, they may create an inaccurate picture of the full range of our interests and complicated personalities. I might not want my students to know that I like to download the music of Richard Strauss, because I don't want them to think of me as the kind of person who likes to download the music of Richard Strauss. I'm not that kind of person at all. Really.

This fear of being judged out of context, I think, is why the subpoenaing of Monica Lewinsky's bookstore receipts struck many people as so invasive; it also helps to explain why nearly all states have passed laws to protect the identities and checkout records of library patrons.[16] Indeed, there was a public outcry in the late 1980s when the FBI disclosed the existence of its "Library Awareness Program," in which the bureau asked librarians around the country to monitor patrons with foreign-sounding names or accents who checked out books on technical and scientific subjects.[17] Similarly, when journalists obtained copies of Robert Bork's video rental records during his Supreme Court confirmation hearings in 1987, anxious legislators responded with the Video Privacy Protection Act, which forbids rental store proprietors from disclosing the viewing habits of their customers. But the act doesn't appear to cover on-line delivery of videos: when you downloaded a video from DIVX (before DIVX went out of business), the National Information Infrastructure Copyright Protection Act protected the right of the copyright owner to demand and record your name before unlocking the movie for you to view.[18] Indeed, the failure of DIVX may be a strong example of consumers reacting badly to an unnecessary loss of privacy. As the sympathy for Judge Bork demonstrates, people don't want to *have* to disclose their viewing preferences in order to watch a movie.

This chapter will argue that new technologies in cyberspace, for all their intrusive potential, can also be exploited as a force for good—to reconstruct the right to read and speak anonymously, not just to erode it. In particular, it will examine ways that law and technology can resurrect in cyberspace the privacy that citizens took for granted in real space in the late eighteenth century. Electronic cash cards, like telephone cards, can enable people like Monica Lewinsky to order books over the Internet with complete anonymity, thus avoiding the risk not only of credit card subpoenas but also of running into Linda Tripp in the bookstore. Digital encryption can ensure that the content of e-mail and computer files is far more secure than it would be if it were locked in John Wilkes's desk. And Web sites can be configured to permit you to disclose selected details about yourself, such as your age, without revealing your actual identity. Virtual blinders, in short, can reduce the risk that personal information revealed in one context is exposed in another. The architecture of cyberspace is political, I will argue, and political choices will determine whether cyberspace embodies values that enhance privacy or values that accelerate its destruction.

The right to speak and read anonymously has played a central role in the history of free expression in America. The Federalist Papers were famously anonymous—"Publius" was a pseudonym for Madison, Hamilton, and Jay; and other Federalist pamphleteers used names such as "An American Citizen," "A Landholder," and "Marcus." (Anti-Federalists responded with "Centinel," "Brutus," and "The Impartial Examiner."[19]) As Justice Clarence Thomas has noted, many of the defining battles over freedom of the press in the Founding era concerned the importance of anonymity in protecting unpopular speakers. The most famous free-speech drama of all involved John Peter Zenger, a printer who refused to reveal the names of anonymous authors who had criticized the royal governor of New York. Unable to prosecute the authors, the governor prosecuted Zenger himself for seditious libel. In an early example of jury nullification, a New York jury refused to convict Zenger in 1735, and the acquittal was widely viewed as a triumph for the related principles of anonymity and freedom of the press.[20]

The right to read anonymously is deeply rooted not only in the First Amendment but also in the constitutional guarantee of freedom from unreasonable searches and seizures. In striking down a law that forbade postal inspectors from delivering Communist literature until the recipients had identified themselves in advance, the Supreme Court emphasized that the law would deter citizens from reading what they pleased.[21] Four years later, in striking down another law making it a crime simply to *possess* obscene materials at home, the Court emphasized the defendant's "right to satisfy his intellectual and emotional needs in the privacy of his own home" as well as "the right to be free from state inquiry into the contents of his library."[22] In other decisions, the Court has protected the right of political organizations not to disclose their membership lists,[23] and to distribute their literature anonymously,[24] as part of the broader rights of free association and free assembly.

In the late eighteenth century, the right to read anonymously was something that citizens took for granted because of inefficiencies in the technologies of monitoring and searching: in order to know what John Wilkes was reading, the king's agents had to break into his house. Now that new methods of online delivery and data collection in cyberspace threaten that anonymity, we face legal as well as technological choices about whether or not to reconstruct it. But First Amendment questions aside, efforts to pass comprehensive legal protections for privacy haven't fared very well in America for a simple reason: although polls about privacy show that a majority of people claim to support it, many of the best-organized interest groups strenuously oppose it. Corporations dislike privacy protections that would restrict their ability to use personal information in marketing schemes. Lobbyists for federal law enforcement are also powerful foes of privacy reform. In fact, the only consistently pro-privacy group in national politics has been the Ruby Ridge wing of the libertarian right, which opposes federal authority in all its forms.

As a result, the politics of privacy tends to be largely reactive, fired by heartstring-tugging anecdotes that capture the public imagination. The murder of actress Rebecca Schaeffer by an obsessive fan who had discovered her address using state driver's license records led Congress to pass the Driver's Privacy Protection Act, which forbids states from releasing personal information such as Social Security numbers, photos, ages, and addresses. Princess Diana's death inspired Senator Dianne Feinstein and Senator Orrin Hatch to propose an antipaparazzi privacy bill, which would hold photographers liable for repeatedly threatening the safety of celebrities for commercial purposes. In 1999 Congress considered several versions of a medical privacy protection act, inspired in part by an article in *The Washington Post* reporting that CVS drugstores and Giant Foods were revealing the private prescription records of patients to a pharmaceutical company. (The company then sent notices to customers encouraging them to refill their prescriptions and offering other treatments for their most private ailments.)[25] But although the issue of medical privacy has great public resonance—most citizens are surprised to learn that their video rental records are more private than their medical records—a powerful coalition of insurance companies, law enforcement interests, and health care businesses has been formed to oppose even the mildest proposals for federal medical privacy legislation.[26]

Press interest in the question of Internet privacy has led to a series of proposals in Congress to protect personal information in cyberspace. And in a sentimental age, privacy issues are often tearfully converted into children's issues. Some of the bills purport to protect children by prohibiting interactive computer services from collecting or disclosing any information about a child without gaining written consent of parents[27] or notifying the child in advance.[28] (But how, precisely, would the child object?) Others would protect personal information more generally, by prohibiting computer services from disclosing information that would identify a subscriber without his or her written consent.[29] Still others would regulate direct marketing and "spam," forbidding advertisers to send unsolicited e-mail,[30] or requiring spam to be labeled as an "advertisement" so that browsers can easily block it.[31] But in the face of opposition from corporate marketers, many of these bills, too, have languished.

Given the precariousness of the politics of privacy, the most encouraging proposals for protecting privacy on-line are technological rather than legal. In 1997, Vice President Al Gore, for example, called for an "electronic bill of rights." In an age when anyone who uses the World Wide Web leaves "a trail of personal data that can be used or abused by others," Gore declared, "You should have the right to choose whether your personal information is disclosed." He announced plans for a new Web site, sponsored by the Federal Trade Commission, that would help Internet users negotiate protocols about privacy before they decide to enter other sites, and so would increase their control over how much personal information they reveal. A more sophisticated privacy protocol, called the Platform for Privacy Preferences, or P3P, can set up a software agent to negotiate with each Web site you visit how much personal information will be collected, and how, precisely, the information will be used. In an electronic mating dance, the Web site makes a proposal about what sort of information it wants to collect. If the proposal is compatible with the privacy preferences that you've specified in advance, it's accepted automatically. If not, P3P can make an alternative proposal on your behalf, or ask the Web site to come up with an alternative. But commercial Web sites may have little incentive to participate in this minuet: I can instruct P3P to tell my bank about my strong taste for privacy, but my bank may respond by refusing to give me a credit card.

Other changes in the architecture of cyberspace could enhance privacy on-line. Julie Cohen of Georgetown Law Center argues that in order to protect constitutional values, copyright management systems should be designed to protect the right to read anonymously. In other words, you should be allowed to download a book without disclosing your ultimate identity to the copyright holder. There are a number of ways that the right to download anonymously could be protected—permitting (rather than forbidding) individuals to pay with untraceable digital cash; prohibiting the collection and disclosure of identifying information for purposes other than those the reader has authorized;[32] or using digital certificates to create mechanisms for pseudonymous downloading.

Within the next few years, individual Internet users could come close to realizing Louis Brandeis and Samuel Warren's ideal: the right of every individual to determine "to what extent his thoughts, sentiments, and emotions shall be communicated to others." Consider e-mail. In chapter 2, I discussed the hazards of a world in which ordinary e-mail sent over a computer network, whether owned by your employer or by AOL, is infinitely searchable and infinitely retrievable, and easily misinterpreted. But already, technologies of encryption exist that make it possible to send "snoop-proof" e-mail messages that no employer can easily invade. Encryption is a way of scrambling the text of a message so that it can't be read unless it's decoded. Anyone can download from the Net an encryption system like the one offered by a company with the modest name of Pretty Good Privacy. This system uses two matched keys, or numerical passwords, one of which is public and the other private. To send you an encrypted e-mail, I encrypt the message with my private key and your public key, which I've learned from a public directory. Only you can decrypt the message using your private key, and you can use my public key to confirm that the message came from me. The problem with encryption technology is that it's still sufficiently complicated to deter casual users with nothing serious to hide. Directories of the public keys aren't widely available, and the whole thing has a whiff of Dungeons & Dragons.

Already, however, user-friendly Web sites are springing up that give you the benefits of encryption without the hassles of having to understand the difference between public and private keys. A site like ZipLip.com, for example, allows you to send encrypted e-mails for free without leaving any records that can be subpoenaed or searched. "Are you at a land phone? It might be more secure if I call you back," said Kon Leong, the president of ZipLip, when I called him on his cell phone. "In my last startup, there was an engineer, and for a long time I wondered how he knew so much about our product. Then I found under his desk that he had a scanner; he was eavesdropping." After sign-ing on to ZipLip, you send your message to the central server, using the same secure link that's used to transmit credit card in-formation. You can sign the e-mail, or send it anonymously, or use a pseudonym, and you can add document attachments or not. You can also seal the e-mail with a prearranged password, so that no one who has access to the in-box except the recipient can read it. ZipLip then sends the recipient an e-mail informing her that there is an e-mail waiting to be read, and after the re-cipient clicks a secure link, she is transported to ZipLip, where she receives the message.

ZipLip destroys all documents and logs on its central server within twenty-four hours, to avoid subpoenas. But the fact that the e-mail is stored in the first place shows that anonymity providers are only as strong as their internal security policies, and many of the architectures of anonymity have weak links. Anonymizer.com, for example, the Web's largest commercial anonymity provider, is a sort of wrapping paper around your Internet browser that makes it possible for you to browse and send e-mail without revealing your identity. Instead of connect-ing directly to a Web site, your Web browser, such as Netscape or Internet Explorer, connects to Anonymizer's proxy computer and asks it to retrieve a Web page. Anonymizer removes all iden-tifying information, forwards the request to the Web site (which knows nothing more than that it's coming from Anonymizer), retrieves the page, and then sends it back to your browser. If your employer is tracking your Internet use, it will know that re-quests are coming from Anonymizer but won't be able to iden-tify which sites you've viewed.

Anonymizer now scrambles all the identifying addresses it receives from individual users, so that it can't trace the identity of its customers even if ordered to by a court. Not long ago, a group of ex-Scientologists used Anonymizer's e-mail system to post an electronic version of the church's sacred documents to an Internet newsgroup. The Church of Scientology, which, unlike most churches, regards its sacred texts as a form of trade secrets and is notorious for suing its critics, issued a subpoena ordering Anonymizer to reveal the identity of the senders. According to Lance Cottrell, the president of Anonymizer, the company was unable to do so. "Our strategy is to be technologically unable to cooperate with subpoenas," Cottrell told me. "We provide full logs of everything we have, and they provide no useful information." By contrast, another anonymity provider, the anon.penet.fi system, was set up by a Finnish computer scientist, Johan Helsingius, who keeps logs that connect the pseudonyms of users to their e-mail addresses. By threatening Helsingius under Finnish law, the Church of Scientology forced him to identify a user who the Church claimed had posted a file stolen from its computers to a Usenet group about Scientology.

As I write, the most advanced technology of anonymity and pseudonymity in cyberspace is being developed by companies such as Zero-Knowledge.com, which is based in Montreal. For a modest fee, you can buy a software package called Freedom, which allows you to create five digital pseudonyms, or "nyms," that you can assign to different activities, from discussing politics to surfing the Web. (Why any of us needs five pseudonyms isn't entirely clear, but the enthusiasm of the privacy idealists is sweet in its way.) Pseudonyms established on ordinary Internet servers are easily traceable: America Online, for example, got into trouble when it revealed that the screen name "boysrch" belonged to a sailor named Timothy McVeigh—a revelation that led the Navy to try to expel McVeigh for homosexuality. Similarly, Raytheon, a defense contractor, sued more than twenty employees in 1999 for posting pseudonymous messages criticizing the company on the Internet; two of the employees resigned in protest after Yahoo revealed their identities in response to a subpoena. (Indeed, Yahoo now has a subpoena group—a cluster of employees who are responsible for tracking down the identity of its subscribers in response to civil subpoenas.) But on the Freedom system, no one, not even Zero-Knowledge itself, can trace your pseudonyms back to your actual identity. "You can trust us because we're not asking you to trust us," says Austin Hill, Zero-Knowledge's twenty-six-year-old president. Hill has a pompadour, a goatee, and a messianic air about his role in vindicating what he considers to be the universal human rights of privacy, free speech, and the possibility of redemption in a world where youthful errors can follow you for the rest of your life. "Twenty years from now, I'm going to be able to talk to my grandkids and say I played an instrumental role in making the world a better place," he says. "As the Blues Brothers say, everyone here feels that we're on a mission from God."

Zero-Knowledge makes traceability very difficult by wrapping the e-mails and Web browsing requests that it receives in layers of encryption, each of which can be opened only by the recipient, and each of which is sent through at least three separate intermediate routers on its way to its final destination. Every message is wrapped like an onion in layers of cryptography, and each router can peel off only one layer of the onion in order to learn the next stop in the path of the message. Because no single router knows both the source of the message and its destination, the identity of the sender and that of the recipient are difficult to link. A request to browse *The New York Times*'s Web page is transmitted in the same way, wrapped in layers of encryption and passed from server to server. Zero-Knowledge assigns pseudonyms using the same technology, and so the company itself can't link the pseudonyms to individual users; if it is subpoenaed, it can only turn over a list of its customers, who can hope for anonymity in numbers. Because Zero-Knowledge is based in Canada, it uses a relatively strong form of encryption—128 bits, which is considered unbreakable in the short term under current technology.[33]

At the moment, the protection offered by anonymous remailers stops at the home computer. The remailers are designed to make it harder for anyone in the outside world to link your pseudonymous identity with your home PC; but they don't protect you if the police or prosecutors get authorization to search your home PC after collecting evidence from other sources. When the PC of a Zero-Knowledge user is searched, says Hill, "there would be full traceability." In this sense, anonymity providers might be seen as doing nothing more than restoring the balance of privacy back to something like the level that existed in John Wilkes's day in the late eighteenth century, when the state had to invade your home and physically search your desk drawers to read your private correspondence.

Whether or not there is a commercial future for anonymity remains to be seen. (As personal computers give way to networked computing, and our personal documents are increasingly stored outside the home, the market for anonymity may increase.) But for ordinary Internet users, authenticity and privacy are more important concerns than complete anonymity, which means that the most hotly contested technical questions of the next decade may concern questions of digital identity. In real space, we authenticate our identities by presenting official documents, such as a driver's license or passport that verifies certain traits about us—our height, weight, and state of residence, for example—while linking those traits to our ultimate identity. In cyberspace, it's possible to create digital certificates that separate our ultimate identity from its authenticating characteristics, greatly increasing each individual's control over how much personal information about himself or herself to reveal. So, for example, if I visit a discussion group about prostate cancer, I can present a certificate that confirms that I'm a man but nothing more. If I want to visit an adult site, I can prove that I'm over eighteen, and a resident of Washington, D.C., without revealing my name.

In their current incarnation, digital certificates, like those provided by VersiSign Digital IDs to authenticate e-mail, tend to store all information about a user in a single place, like a driver's license. But there is no reason that states, or private certificate authorities, couldn't issue digital driver's licenses that unbundle the different traits of my identity into different digital certificates. Stored as separate computer files, each certificate could authenticate my age, hair color, state of residence, gender, and so forth. Web sites could then be configured to demand no more information than they need in order to decide whether or not to give me access.

There is something exhilarating about the disaggregation of identity in cyberspace. In a world of ethnic balkanization and identity politics, in which reasoned deliberation seems difficult because, we're sometimes told, none of us can transcend our racially and sexually and economically determined perspectives, the promise of selective anonymity seems liberating. There's no possibility of being victimized by stereotypes when you speak anonymously; on the Web, people can be judged by what they say rather than who they are. But the disaggregation of identity is a double-edged sword. "Ordinarily the very purpose of anonymity," Justice Antonin Scalia has argued, is to "facilitate wrong by eliminating accountability."[34] In addition to protecting heroic Chinese dissidents posting to a newsgroup about democracy, anonymous remailers can protect child pornographers, electronic thieves, stalkers, harassers, and libelers.

As digital certificates and digital cash become increasingly widespread, and as we conduct more of our lives on-line, the increase in privacy and selective anonymity will make it harder to detect and punish crime. How can the legitimate interests of law enforcement be preserved in cyberspace? The most heavy-handed solution would be to outlaw anonymous remailers entirely—as the Georgia legislature tried to do in 1996 when it passed a bill criminalizing certain types of anonymous and pseudonymous speech on the Internet—prohibiting, in particular, the transmission of data accompanied by names that "falsely identify" the sender.[35] But such a law would be unlikely to survive constitutional scrutiny: in enjoining the Georgia law in 1997, a federal district court emphasized that anonymous speech is protected by the First Amendment and that "the identity of the speaker is no different from other components of [a] document's contents that the author is free to include or exclude."[36] "False identification," the court concluded, might be a legitimate tool "to avoid social ostracism, to prevent discrimination and harassment, and to protect privacy."[37]

There are less clumsy ways that the government might try to preserve its ability to prosecute crime and fraud in cyberspace. Congress might require, for example, that the link between your real-world identities and your cyberidentities be ultimately traceable by the state. It could require, in theory, that an encrypted fingerprint or retinal scan be attached to every transaction in cyberspace before an individual could have access to the Web, although one hopes that public opposition would make this politically infeasible. The fingerprint could be encrypted with the government's public key, so that state agents, but no one else, would be able to access your real identity if they had good reason to suspect you of a crime. A requirement that all identities be put in escrow might be upheld by the Supreme Court: in 1999, Justice Ruth Bader Ginsburg held for the Court that states couldn't require signature collectors to wear name tags when they tried to gather support for petitions to place initiatives on the ballot. But Ginsburg said that the signature collectors could be required, when they turned in their signatures, to include an affidavit with a notarized copy of their name, address, and signature. The notarized information would be available to law enforcement officials if there were any allegations of election fraud, but would not be disclosed to the ordinary citizens whose votes were being solicited.[38]

Under what circumstances should identifying information be available to law enforcement? In 1999 Congress proposed to relax export controls over encryption, but at the same time to make it a crime to fail to decrypt encrypted information, and turn over a readable plain-text version to the government, when ordered to do so by a court. If adopted by Congress without stronger procedural protections, the proposal might undermine the central achievement of encryption technology, which has restored some of the protection for private papers that citizens enjoyed in John Wilkes's day. I argued in chapter 1 that judges in the eighteenth and nineteenth centuries were willing to balance the need for private papers against the seriousness of the crime, and were far more willing to compel the production of diaries and letters in cases involving mass murder than in civil and white-collar criminal investigations. If eighteenth-century values are to be translated into the age of cyberspace, Congress today could require judges or special privacy masters to engage in the same constitutional balancing test, issuing court orders to decrypt secret documents only in cases where individual suspicion is high and the crime being investigated is very serious indeed.

Another way of ensuring that citizens today have the same security of their electronic papers and effects that citizens in the eighteenth century took for granted is to create new property and contract rights in personal information. Because the Fourth Amendment is rooted in notions of private property, as I discussed in chapter 2, judges have had trouble regulating forms of electronic surveillance that don't clearly invade property rights. But a new generation of scholars is focusing on the way that property and contract rules could protect the privacy of electronic information. Congress or state legislatures could give citizens property rights in their personal data and could prohibit the data from being used or sold without the property holder's permission, or consumer contracts between buyers and sellers could determine the terms of disclosure.

There are, however, limits to the degree that personal data can be conceived of as private property, especially because property rights can be sold. Some people don't care about privacy until they have something to hide, and there's no reason to believe that consumers wouldn't voluntarily transfer property rights in their personal data to commercial Web sites in exchange for product discounts and other conveniences. This might leave consumers worse off in the long run: someone who trades the property rights in his on-line purchases to a drugstore when he is healthy would have no way of recovering his privacy once he is diagnosed with an embarrassing disease. For this reason, contract and tort rights, which allow individuals to sue private marketers for breaching confidences and nondisclosure agreements, may be a more effective way of protecting the privacy of click-stream data.[39]

In the age of John Wilkes, judges were willing to make it more difficult to enforce sedition and blasphemy laws because of the dangers of exposing innocent books and papers in the course of searching for illegal books and papers. A similar sensitivity to the risks of overintrusive searches should inform the way we address the most politically contested question about privacy in cyberspace today: how to protect children from cyberpornography. In real space, it's hard for a fifteen-year-old to buy pornography anonymously; he could walk into a store wearing dark glasses and a fake mustache, but the disguise would be unlikely to fool a clerk. Moreover, in real space, the Supreme Court has said that it's permissible to impose liability on pornography distributors who fail to screen out underage customers: in 1965, Sam Ginsberg and his wife were running a luncheonette in Bellmore, Long Island, when a sixteen-year-old named Richard Coray walked in to buy some pinup magazines. Without asking for identification, Ginsberg sold him the magazines, whereupon he was convicted of violating a New York law that made it illegal to sell material "harmful to minors" to a child under seventeen, even if the material wouldn't be considered obscene if it were sold to adults. In upholding Ginsberg's conviction, the Supreme Court stressed that states can forbid the sale of material to children that they couldn't deny to adults, and they can require distributors to act as intermediaries, checking IDs to distinguish adults from children.[40]

In cyberspace, it's easier for most children to get pornography than it is in real space: anonymity is pervasive, and much of the material is free. In 1995, as public access to cyberspace began to explode, the *Georgetown Law Journal* published a flawed study that claimed that "83.5% of all images posted on the Usenet are pornographic."[41] The study, conducted by Marty Rimm, a thirty-year-old undergraduate at Carnegie Mellon University, was the basis for a lurid cover story in *Time* magazine, featuring a horrified tot with an Edvard Munch expression staring at a computer screen.[42] Like the *Time* story, Rimm's data were attacked on the Internet as tendentious and unreliable. Rimm's credibility collapsed entirely when he acknowledged that he had used the data from his Carnegie Mellon study for what was to have been an illustrated paperback entitled "The Pornographer's Handbook: How to Exploit Women, Dupe Men, and Make Lots of Money" and designed to help operators of adult bulletin boards market their wares more effectively. Nevertheless, the scandal created the perception of public pressure for the regulation of cybersmut, and Congress responded in 1996 by passing the Communications Decency Act. The CDA made it a felony to transmit "indecent" or "patently offensive" material on the Internet in a way that made it available to anyone under eighteen. But it also said that indecent speakers on the Internet would have a defense to prosecution if they took "reasonable, effective, and appropriate actions" to restrict access to minors, by "requiring use of a verified credit card, debit account, adult access code, or adult personal identification number."

In 1997, the Supreme Court struck down the CDA as a violation of the First Amendment, stressing that Congress had no power to regulate the vaguely defined category of "indecent" speech (except in the context of television broadcasting).[43] The definition of indecency was so broad, the Court noted, that "a parent who sent his 17-year old-college freshman information on birth control via e-mail could be incarcerated even though neither he, his child, nor anyone in their home community, found the material 'indecent' or 'patently offensive,' if the college town's community thought otherwise." The Supreme Court also noted that, at the time of the trial, existing technology didn't permit a sender to prevent minors from obtaining access to speech without also denying access to adults. Because the CDA covered indecent speech transmitted through e-mail, newsgroups, and chat rooms, as well as postings to the Web, the Court found that it would be prohibitively expensive to expect noncommercial speakers to verify that their users are adults, by checking credit cards, for example, or digital certificates. In a concurring opinion, however, Justice Sandra Day O'Connor concluded that the CDA was a well-intentioned, if flawed, effort to create zones of pornography on the Internet that protected children from material that was appropriate for adults, and she suggested that a better-designed "zoning" law might pass constitutional muster as technologies for the verification of identity evolved.[44]

There is, however, a zoning solution to the pornography problem that could protect children without restricting the access of adults—a solution based on the kids' ID rather than the adults' ID. The government could require the manufacturers of browsers, such as Microsoft or Netscape, to allow individual users to identify themselves as minors in their browser profiles. (If children and adults used the same machine, adult profiles could be set up to require a password.)[45] Alternatively, Netscape and Microsoft could require that individual users prove, through digital certificates, whether they are children or adults.[46] Adult browsers would give users access to the entire Web; kids' browsers would provide access only to sites that weren't blocked for children. Commercial pornographers could be required to deny access to anyone who tried to sign on from a kids' browser.

The proposal to zone the Internet with kids' browsers is less threatening to free speech than proposals to adopt voluntary rating systems. An organization called the World Wide Web Consortium, for example, has developed a program called the Platform for Internet Content Selection, or PICS, to allow parents to restrict what their children view on the Web.[47] PICS is a rating and filtering technology, like the V-chip, that permits private organizations, from the Christian Coalition to the American Civil Liberties Union, to set up private rating systems for any "PICS-compatible" document that is posted on-line. Individual users can then choose the rating system that best reflects their own values, and any material that offends them will be blocked from their homes. Although PICS promises to allow individual users to exercise perfect choice about what comes into their homes, it raises a troubling new set of problems. Lawrence Lessig, for example, suggests that "PICS is the devil," from a free speech perspective, because it allows censorship at any point on the chain of distribution. Countries like China and Singapore can decide what kind of speech they want to make available to their workers, and impose draconian restrictions from above. In the long run, Lessig suggests, PICS will suppress more speech than a kids' identification system would, because it will allow those who control access to computer networks to filter out uncongenial ideas as well as hostile or offensive speech.

The efforts by universities, corporations, and public employers to adopt Internet filters in order to avoid liability for sexual harassment, which I discussed in chapter 2, have vindicated Lessig's fears. And by creating strong incentives to adopt Web filters, harassment law has inhibited the revolutionary potential of the Internet, which has the ability to increase individuals' control over how much personal information they disclose or receive. Rohan Samarajiva defines privacy as "the capability to explicitly or implicitly negotiate boundary conditions of social relations,"[48] and the Internet, properly designed, puts this capability squarely in the hands of individuals rather than intermediaries. Indeed, in *The Control Revolution,* Andrew Shapiro argues that allowing individuals to bypass information middlemen— such as stockbrokers, book retailers, and newspaper publishers— is the Internet's most distinctive feature. Instead of relying on editors and publishers to decide what we should read and hear, the Internet makes individuals producers as well as consumers of information, armed with the ability to decide for ourselves what is likely to inform or offend us. So, for example, news personalization services, such as those offered by *Slate* magazine, are already making it possible to custom-design your daily newspaper so that it includes only the columnists, reporters, and subjects that interest you.

This increase in individual control over the inflow and outflow of information, which allows each of us to construct our own filters to determine what we will see and hear, has costs as well as benefits: Shapiro worries that the technologies that allow individuals to exercise a "freedom from speech" will lead to a growing "privatization of experience"[49] that can descend into solipsism: someone who has a moral aversion to homosexuality, for example, may be able to construct a filtered universe in cyberspace that ensures she is never confronted with articles that might challenge her preconceptions, while in real space, this sort of self-protective narrowcasting is harder to achieve. At the same time, the decline of editorial middlemen makes possible the rise of cybergossips like Matt Drudge who publish rumors without purporting to vouch for their truth. By printing the stories that responsible newspapers are hesitant to print, Drudge increases the pressure on the mainstream press to run follow-up stories about the rumors that it had initially suppressed. In Brandeis and Warren's day, the greatest technological threat to the privacy of public figures was instant photographs published in tabloid newspapers. In the age of the Internet, by contrast, the opportunities for broadcasting gossip are infinitely more powerful, and what used to be proclaimed from the housetops is now discussed in chat rooms that are accessible in living rooms around the globe.

In some ways, the new electronic media look less like the twentieth-century broadcast networks or the nineteenth-century tabloid newspapers, and more like the interactive eighteenth-century printing presses that were at the core of the original understanding of the First Amendment.[50] At the time of the framing of the American constitution, printers, who were also publishers, delegated the task of news gathering to their readers, who also served as reporters. Presses were small, and barriers to entry were few: 450 newspapers started up between 1783 and 1801. The pamphlet presses of the eighteenth century are similar in some ways to the electronic bulletin boards of today. But whereas eighteenth-century pamphlets had a relatively restricted audience, Internet gossip is globally accessible and hard to delete, which makes it more likely that personal information published in a chat room may be taken out of context.

The Internet has decreased the cost of and increased the opportunities for publishing false and misleading information. But because it also increases the opportunities for falsehoods to be corrected, some cyberexceptionalists have argued that, in cyberspace, there is no need for a law of libel. For example, in *Cyber Rights: Defending Free Speech in a Digital Age,* Mike Godwin points to three differences between cyberspace and real space to support his conclusion that libel law is unnecessary on the Net. First, Godwin argues, the Internet has changed the nature of public figures. Since anyone with a modem and a keyboard can now "thrust himself into the vortex of [a] public issue," to borrow Justice Lewis F. Powell's phrase in the Gertz case, many more people now count as public figures, or limited public figures, for libel purposes. Second, Godwin argues, the Net is reducing and perhaps even erasing the imbalance of power between mass media and private individuals that made libel law necessary in the first place. Because private figures now have the same broad access to the Internet to correct false statements of fact about themselves, the justification for giving private figures more protection against libel than public figures is evaporating.[51]

Finally, Godwin argues, Internet service providers, such as AOL and CompuServe, are more like bookstores than like traditional publishers, and therefore they shouldn't be held liable for falsehoods published on their watch.[52] Congress seemed to accept this argument, in a portion of the Communications Decency Act that the Supreme Court *didn't* strike down, which gave Internet service providers broad immunity from liability for defamatory statements made by others.[53] This section was designed "to maintain the robust nature of Internet communication"[54] and to encourage free speech by preventing service providers from monitoring and restricting—out of fear of liability—the messages that can be posted on-line. Indeed, when Matt Drudge reported false rumors that Sidney Blumenthal, the Clinton aide, beat his wife, Blumenthal sued not only Drudge but also AOL, which had hired Drudge as a featured columnist. A federal judge, however, dismissed AOL from the suit. Congress, the judge noted, had made a decision not to hold Internet service providers responsible for the speech of columnists whom they pay and promote. Congress's logic was that AOL should be able to monitor the speech that occurs in its chat rooms and bulletin boards, without fear that the monitoring itself will be viewed as a form of editorial control that converts AOL from a bookstore into a publisher.[55]

Although Congress has ensured that there's less of a financial incentive to file libel suits in cyberspace, this doesn't mean that there's no role at all for libel law on the Internet. It's not clear, first of all, that a private citizen who is the victim of libel in cyberspace really does have the ability to correct the misimpression: if Matt Drudge, in a fit of pique, published nasty rumors about his dry cleaner, the dry cleaner could establish a Web page to set the record straight. But because Drudge has such a wide circulation, far fewer people would see the dry cleaner's Web page than saw the original rumor.

Most citizens will never be attacked by Matt Drudge, but many of us may find ourselves the object of gossip in Internet chat rooms. Unlike gossip among neighbors, moreover, the gossip published in cyberspace is often exchanged by individuals who don't know the people they are gossiping with or about; lacking the restraint that accompanies face-to-face encounters, the discourse can be venomous as well as inaccurate. For example, on-line magazines, such as *Slate* and *Salon,* have established chat rooms for their readers to comment on the articles they have just read. The postings often take the form of rumors and unkind speculation not only about established journalists, but also about private citizens who have ventured into cyberspace journalism for the first time. Unlike gossip around the office watercooler, moreover, gossip in cyberspace can be retrieved anywhere on earth and never goes away.

Ordinary newspapers have editors who screen out inaccurate or tasteless material. Because *The New York Times* has careful editors and reporters with a reputation for accuracy, you're more likely to believe what you read in the *Times* than what you encountered in my own magazine, *The New Republic,* during the period when a twenty-five-year-old reporter was discovered to have made up a substantial part of his corpus. The young man, Stephen Glass, not only fabricated characters and invented scenes but also invented fact-checkers' notes and Web sites in order to trick the fact-checkers who were responsible for correcting him. But *The New Republic,* as soon as it discovered that it had a journalistic con man in its midst, fired Mr. Glass, who was banished from the journalistic community. This signaling and stigmatizing function also served the broader purposes of libel law. Robert Post reminds us that the reason courts impose liability for false statements of facts isn't merely to correct inaccuracy but also to reintegrate the victim into the community of respectable citizens, to publicly acknowledge the wrong, and to express it in a social, communal way.[56] The traditional media, without legal regulations, seem to perform this self-policing mechanism quite well.

In cyberspace, however, the incentives are different. As a self-publisher, Drudge is Drudge whether you access him from AOL or from Netscape, and he has no editors to filter or check him. (In response to criticism from President Clinton's press secretary, Drudge once replied, "What is Mike McCurry going to do, call my boss?"[57]) And the self-contained communities of Internet gossips have even less incentive to measure their words. Law is too clumsy an instrument to protect individuals from false statements in cyberspace except in the most egregious cases, and there would be serious risks to freedom of speech if individual cybergossips had to fear being bankrupted by libel judgments every time they took to the keyboard. Nevertheless, if one of the purposes of libel law is to perform the reintegrative function that I've described, it's not clear that cybergossip should be completely unregulated. In a truly extreme case involving cyberlibel against a private figure, perhaps a declaratory judgment that wrongdoing has occurred (rather than ruinous punitive damages) would be an appropriate way of acknowledging the victim's injury, and correcting the misimpression, without inhibiting the free flow of ideas.

Some of these lessons about libel in cyberspace can be applied, I think, to sexual harassment in cyberspace as well. When it comes to targeted acts of harassment—hostile e-mail or spam sent from one person to another—technology currently offers more effective protections than law. It's easy to design a commercial e-mail service that can block incoming messages from designated individuals: when you receive an offensive message from a sender and decide you'd rather not receive another, you can add the sender's address to an electronic kill file, so that all future messages from it will be deleted automatically.[58] Even anonymous e-mailing services can block messages from designated senders. "I may have the freedom to be private, but you have the freedom to say 'I don't want to talk to that particular person,' " says Austin Hill of Zero-Knowledge. "If you're John Smith being harassed by Superman, you can go to 'abuse' at Freedom.net and say, 'I'd like to block any e-mail coming from Superman,' or say, 'I want to block e-mail from everyone but Superman.' " Indeed, in real space, courts have upheld laws that prohibit mailers from sending material to people who have expressly asked not to receive it; there is no reason that the same screening function can't be achieved technologically in cyberspace.[59]

What about offensive speech or true but embarrassing facts broadcast in chat groups? In the interest of maximizing privacy as well as freedom of speech, I've argued that liability for creating a hostile environment in real space should be imposed on individual speakers, rather than on employers, and in the absence of tangible job consequences, it should take the form of suits for invasions of privacy rather than for gender discrimination. In cyberspace this model makes even more sense, as the opportunities for widespread circulation of gossip increase. Consider, for example, the remarkable case of Santa Rosa Junior College, where the faculty advisor to the college newspaper set up a virtual discussion group for women only, at the request of female students who said they felt freer to discuss women's issues in isolation from men. In the interest of fairness, the advisor also set up an all-male discussion group, and a coed one as well. The trouble began when a female student protested what she thought was a sexist advertisement in the student newspaper, depicting the rear end of a woman in a bikini. Some of the men in the male conference posted "anatomically explicit and sexually derogatory"[60] comments about the protestor, and one man also posted vulgar comments about his ex-girlfriend, a staff member on the student newspaper. Breaking the pledge of confidentiality that was a condition for admission to the group, a member of the all-male group told the two women about the profane comments. They immediately complained to the faculty advisor, who responded by shutting down the all-male and all-female discussion groups.[61] Not satisfied, the two women filed a complaint with the Office of Civil Rights at the U.S. Department of Education, which concluded that the women had probably suffered from sexual harassment because the comments appeared to be "so severe and pervasive as to create a hostile educational environment on the basis of sex."[62] The college eventually settled with the two women, paying them fifteen thousand dollars each in exchange for being released from legal liability.[63] But the Office of Civil Rights then took it upon itself to propose a set of rules for future computer discussion groups at Santa Rosa, which the college adopted, prohibiting the following "harassing conduct": "epithets, slurs, negative stereotyping, or threatening, intimidating, or hostile acts, that relate to race, color, national origin, gender, or disability." The policy continues: "This includes acts that purport to be 'jokes' or 'pranks,' but that are hostile or demeaning."[64]

This story is distressing on several levels. Whatever the two women students experienced once they learned that men were gossiping about them on a private discussion group, it wasn't gender discrimination. Although offensive, the comments ridiculing the woman who protested to the student newspaper were a classic example of what should be protected as political speech. By contrast, the woman whose ex-boyfriend posted unflattering comments about her private life was the victim not of sexual harassment but, arguably, of an invasion of privacy. Under different circumstances, she could have sued her former boyfriend under the branch of invasion of privacy law that prohibits the publication of private facts. To succeed on a public disclosure claim, however, the plaintiff has to prove that the publicized information "contains highly intimate or embarrassing facts about a person's private affairs, such that its publication would be highly objectionable to a person of ordinary sensibilities"; that the information was "communicated to the public at large," not simply to "a small group of persons"; and that the information publicized is not "of legitimate concern to the public."[65] The single-sex conferences were, by definition, private and limited to a small group of students. "Private conferences, by definition, are *not* public," the conditions for participation specified. "It is a violation of the rules of these private conferences to show messages or discuss their contents (other than a general summary) with a non-member of that conference. . . . If you gain admittance, then later violate that rule, you will be summarily thrown out of the conference."[66] Because of the limited nature of the group, it's unlikely that the young woman whose privacy was violated could make out a convincing legal claim. But in a truly egregious case of sexual gossip that was circulated more broadly, an invasion of privacy suit might indeed be appropriate.

The privacy invasions suffered by the Harvard dean, the Santa Rosa college students, and the Internet user who receives targeted ads on the basis of click-stream data show that as the personal information of private citizens is increasingly recorded in cyberspace, there is a growing danger that it will be taken out of context. But although the technology exists to reconstruct the protections for private papers that have been eroded in real space, there is no evidence that the political will exists to use it. The path of privacy in the hundred years since Warren and Brandeis published their famous article shows that when courts and citizens greet new technologies of surveillance and monitoring with passivity and resignation, the boundaries of privacy will continue to erode. The future of privacy will be determined not by the inherent nature of the Internet, but by social choices about how much privacy we as a society think it is reasonable to demand. And failure to choose means that the slow erosion of protections for private papers and personal information that began at the beginning of the twentieth century will be consummated in the twenty-first century rather than stopped short. Will we be passive in the face of technological determinism, or do we have the vision to insist on rebuilding the privacy we have lost?

Epilogue

———

What Is Privacy
Good For?

D o Americans have more or less privacy in the age of cyberspace than we did a generation ago? I began by arguing
that we have fewer constitutional protections against the surveillance and extraction of personal information by state authorities than we did in the eighteenth and nineteenth centuries,
thanks to a series of relatively recent decisions by the Supreme
Court. But if privacy is defined as the ability to protect ourselves
from being judged out of context by controlling the conditions
under which we reveal personal information to others, then protection from state authorities is only one aspect of privacy, and,
for most citizens, not the most salient one. Indeed, John Stuart
Mill insisted that the stifling pressures of social conformity pose
a greater threat to human individuality than legal pressures do.
The mandates of society, Mill wrote, can constitute "a social
tyranny more formidable than many kinds of political oppres-
sion, since, though not usually upheld by extreme penalties, it
leaves fewer means of escape, penetrating much more deeply
into the details of life, and enslaving the soul itself."[1] Mill con-

cluded that "protection, therefore, against the tyranny of the magistrate is not enough: there needs protection also against the tyranny of the prevailing opinion and feeling." In this regard, at the beginning of the twenty-first century, law and technology have increased privacy in some respects and decreased it in others. Let's review the balance sheet.

It's easier to be relatively anonymous in the age of cyberspace, because so many of the ordinary transactions of life can be conducted without leaving the house. From buying clothes and groceries to buying pornography, shopping is increasingly taking place on-line. Watching movies and browsing in a bookstore—formerly public activities—can now occur in the quasi privacy of cyberspace. And because one can buy lingerie or flowers or prescriptions for embarrassing diseases without the risk of running into nosy neighbors at the corner store, the ability of those neighbors to enforce social norms through face-to-face interactions is diminished, and privacy, in the sense that Mill conceived it, is accordingly enhanced.

It's true, as I argued in the last chapter, that transactions in cyberspace tend to generate detailed electronic footprints that expose our tastes and preferences to the operators of Web sites, who can then sell the information to private marketers. But to the frustration of professional privacy advocates, Americans don't always seem terribly concerned about the possible misuse of click-stream data. Many of us use credit cards for the most intimate on-line purchases. We willingly accept cookies, and we don't take the time to cover our electronic tracks with cumbersome anonymity providers, such as Zero-Knowledge. For citizens engaged in day-to-day transactions, convenience often outweighs the theoretical possibility that personal information may be disclosed to strangers. By contrast, citizens do take care to conceal information that is inherently embarrassing: an old-fashioned pharmacy in Georgetown is said to do a brisk business selling antidepressants, Viagra, and other intimate medications to Washington dignitaries, because of its policy of dispensing all prescriptions by hand and not keeping computer records.

Georg Simmel noted that citizens are sometimes more likely to confide in anonymous strangers than in colleagues and friends. "[T]he stranger who moves on," he observes, "often receives the most surprising openness—confidences which sometimes have the character of a confessional and which would be carefully withheld from a more closely related person."[2] This phenomenon of the stranger will be familiar to anyone who has overheard two seatmates thrown together by chance on an airplane or a train loudly sharing intimate disclosures about their most embarrassing secrets. Confessions to strangers are costless, precisely because the social disapproval of strangers can be ignored, unlike the social disapproval of those whom we encounter on a daily basis. Our willingness to reveal personal details about ourselves on the Internet is a technological manifestation of the phenomenon of the stranger. There is no reason for most of us to fear the disclosure of disaggregated bits of personal information to faceless Web sites, because those Web sites, by and large, have no motive or opportunity to collect the data into a personal narrative that could be disclosed to anyone who actually knows us. By contrast, as the reaction to the Double-Click profiles shows, people don't want their browsing habits collected in personally identifiable dossiers, because those dossiers can be bought or subpoenaed by employers, insurance companies, divorcing spouses, and others who have the ability to affect our lives in profound ways. This is why it is perfectly rational for citizens in the age of the Internet to feel more isolated and more exposed at the same time. We fear the loss of anonymity in cyberspace precisely because we are spending more time browsing alone and less time with friends, family, and neighbors who can put our speech and conduct in a broader context.

In cyberspace, the possibilities that personal information may be taken out of context continue to increase. As electronic footprints are collected in centralized databases, they become more accessible to colleagues as well as strangers. A few decades ago, if I bought a house in Washington, D.C., the deed would be inconveniently buried at the municipal courthouse, available only to those who took the time to dig it out. Today, the purchase price is recorded on-line and likely to be published in *The Washington Post,* which keeps a running tally of real-estate transactions—a disclosure that strikes many home buyers as a gratuitous invasion of privacy. If I taught at a state university, my salary, too, might be available on-line, although many academics think that salaries should be a private matter. And if, in a moment of youthful enthusiasm, I once posted intemperate comments to an Internet newsgroup, those comments are likely to be recorded on a Web service like Dejanews, where they can be retrieved years later simply by typing my name into a popular search engine.

In certain social circles, it is not uncommon for prospective romantic partners to perform background checks on each other, scouring the Internet and other electronic search engines, such as Nexis, for as much personal information as possible before going out on dates. And a search like this can be a deal-breaker: a friend of mine, after being set up on a blind date, ran an Internet search and discovered that her prospective partner had been described in an article for an on-line magazine as one of the ten worst dates of all time; the article included intimate details about his sexual equipment and performance that she was unable to banish from her mind during their first—and only—dinner. These are the sort of details, of course, that friends often exchange in informal gossip networks. The difference now is that the most intimate personal information is recorded indelibly and can be retrieved with chilling efficiency by strangers around the globe.

Throughout this book, I have discussed the danger of being judged, fairly or unfairly, on the basis of isolated bits of personal information that are taken out of context. And to the degree that cyberspace has altered our conceptions of time and geographic space, it increases this risk. Shielded by virtual anonymity, people from different parts of the country who meet in chat rooms can be tempted to reveal personal details about themselves too quickly, and are often disappointed to discover, when a face-to-face meeting is finally arranged, that they have constructed a false impression of one another. Indeed, the risk of mistaking information for knowledge in cyberspace is not limited to personal interactions. In a world where Shakespeare is available on the Internet, it's easy to allow search engines to replace the work of reading and mastering the plays themselves. Plugging the word "reputation" into a Shakespeare browser will reveal a series of quotations that can be used out of context, but this is a cheap substitute for the knowledge that can emerge only through slow, solitary study.

Other changes in media technology have increased the risk of mistaking information for knowledge. Because people who are overloaded with information have short attention spans, the explosion of television channels in a digital world may increase consumer choice while making it more difficult for citizens to focus on complicated questions of common concern. Moreover, the proliferation of sound bites in a twenty-four-hour news cycle may misinform citizens rather than educate them, substituting isolated bits of information for genuine understanding.[3] Citing this danger, Justice Antonin Scalia told a class at Harvard Law School that he opposes allowing television cameras into the Supreme Court, because of his fear that sensational snippets of the oral arguments would be taken out of context and excerpted on the evening news. If the arguments were broadcast in their entirety, Scalia said, he would abandon his opposition to cameras in the courtroom.[4] Privacy for the justices, Scalia convincingly suggests, may be a way of increasing understanding of the Court, rather than thwarting it, by forcing citizens to focus on the Court's written opinions rather than on misrepresentative flashes of drama.

The danger of mistaking information for knowledge is illustrated by the phenomenon of the celebrity. When we know a series of intimate details about a famous person—his mannerisms, his love life, his quirks of speech and dress—from celebrity profiles or from watching television, we may be lured into a false sense of intimacy with him. A self-possessed private citizen has an inviolate personality, surrounded by boundaries of reserve that can't be penetrated casually by strangers. A celebrity, by contrast, has an interactive personality: people feel free to approach Sam Donaldson on the street. But the feelings of intimacy that celebrity generates are either misleading—we don't really *know* a television celebrity, even though he may appear every night in our living room—or a sign of self-violation: when a celebrity leads so much of his life in public that nothing is held back for genuine intimates, then he becomes a buffoonish self-caricature, almost literally a talking head, devoid of the

individuality, texture, and boundaries of reserve that character-
ize a genuinely self-possessed personality.

In an age when disaggregated personal information is cen-
trally collected, widely accessible, and permanently retrievable,
private citizens run the risk of being treated like celebrities in the
worst sense, vilified rather than celebrated on the basis of isolated characteristics. They risk suffering from the phenomenon
of the synecdoche, experienced by Clarence Thomas, where the
part comes to stand for the whole. But perhaps terms like
"synecdoche" and "disaggregated personal information" are
too fancy; I am really making an argument about cyberspace
and gossip. In traditional, rural societies, characterized by what
Ferdinand Tönnies called *gemeinschaft,* or community,[5] social
norms were enforced primarily through gossip. By contrast, in
modern, urban environments, characterized by what Tönnies
called *gesellschaft,* or society, a thriving culture of oral gossip is
harder to sustain: intimate information is more difficult to collect and to transmit because neighbors encounter one another
less frequently. Gossip can be a powerful form of behavioral
regulation: when neighbors gossip aggressively about one another's intimate activities, those who transgress community
norms will soon feel the effects of social disapproval, even if
they are not directly confronted about their indiscretions. Nevertheless, in a closely bounded community, where gossip is transmitted orally, there is less of a danger of mistaking information
for knowledge. Because all of the relevant parties know each
other well from close personal observation, individual transgressions can be weighed against the broader picture of an individual's personality.

Oral gossip, Ferdinand Schoeman notes, is a remarkably
flexible way of enforcing social norms while still respecting the

privacy of the individuals who are being gossiped about. As long as gossip remains in the oral realm, it is not treated as fully public information. The circulation of oral gossip is relatively restricted, and the individuals who are being gossiped about can adjust their behavior to conform with social norms without being formally confronted by those who are discussing them behind their backs.[6] The community can hold the individual accountable for his behavior; but because his public face isn't directly threatened, the individual can correct his misbehavior without feeling that his social standing is assaulted at its core. Or he can argue that he has been misrepresented, and try to correct the misperception, with less fear that the unfair gossip is indelible and will return to haunt him in the future.

In his famous essay on reputation, E. L. Godkin elaborated on this distinction between oral and written gossip. As long as gossip was oral, and confined to acquaintances rather than strangers, Godkin wrote in 1890, its objects were often spared the mortification of knowing they were being gossiped about. "A man seldom heard of oral gossip about him which simply made him ridiculous, or trespassed on his lawful privacy, but made no positive attack upon his reputation," Godkin declared. "His peace and comfort were, therefore, but slightly affected by it." Citing Godkin's essay, Brandeis and Warren concluded that the law should not ordinarily grant any redress for the invasion of privacy by oral, as opposed to written, communications about private matters, because the resulting injury is usually so trivial that, in the interests of free speech, it might as well be disregarded. Today, a number of courts have agreed with this distinction.[7]

Cyberspace, however, has blurred the distinction between oral and written gossip and between public and private facts. In real space, gossip published in the *National Enquirer* can be discounted, not necessarily because it isn't true, but because publication in a newspaper that is recognized as being devoted to gossip doesn't convert something from private to public information in a normative sense.[8] Like oral gossip, the circulation of the *National Enquirer* is restricted to the community of gossip connoisseurs. Similarly, information that is published in a cyber-gossip column like Matt Drudge's doesn't have to be dignified with a public response. (Indeed, when presidential press secretaries are asked to respond to stories published by Drudge, they can get away with statements like "I won't dignify that with a response.") But once gossip has been reported on the Internet, its quasi publication gives respectable newspapers, such as *The Washington Post,* an excuse to report it as well, often in the thin disguise of a "media watch" column describing the pressure that Drudge's decision to report a story exerted on other media outlets to report it as well.[9] And unlike information in the *National Enquirer,* information that is published in *The Washington Post* must be treated as public in the fullest sense, because *The Washington Post*'s editorial decisions define the category of information that is deemed to be worthy of public concern.

Defenders of Drudge such as Michael Kinsley argue that Drudge's standards of accuracy are only slightly lower than those of the mainstream media, and that "there is something slightly elitist about the attitude that we journalists can be trusted to evaluate such rumors appropriately but that our readers and viewers cannot."[10] But to focus on Drudge's accuracy misses the point. The problem with Drudge is not, by and large, that the gossip he reports is inaccurate, but that he is part of a hypocritical and brutalizing dynamic by which information that should be treated as private is converted into information that must be treated as public.

Even if gossip in cyberspace never bubbles up into the traditional press, it is more widely broadcast and more easily misinterpreted than it is in real space, resurrecting all of the stifling intimacy of a traditional society without the redeeming promise of being judged in context. The fact that gossip in cyberspace is recorded, permanently retrievable, and globally accessible increases the risk that an individual's public face will be threatened by past indiscretions. Gossip published on an Internet chat group may, in the short run, reach an audience that is no bigger than gossip over the back fence in a small town. But because Internet gossip, unlike individual memories, never fades, it can be resurrected in the future by those who don't know the individual in question, and thus are unable to put the information in a larger context. And unlike gossip in a small town, Internet gossip is hard to answer, because its potential audience is anonymous and unbounded.

The ubiquity of computer databases contributes to this confusion of information with knowledge, of narrative truth with historical truth: a youthful indiscretion, committed and written about at one stage in an individual's career, may be resurrected by journalists or employers for years to come. In such a world, it is easy for individuals to be victimized by the reductionist fallacy that the worst truth about them is also the most important truth. And because gossip that is archived never goes away, it may be hard for individuals to change their behavior, and to adjust their public face, in a way that frees them from the burden of their past mistakes.

There are few experiences more harrowing than being described: to be described is to be narrowed and simplified and judged out of context. And there are few acts more aggressive than describing someone else. I've found writing profiles to be more unsettling than any other form of journalism, because presuming to judge someone else's character, and to present it as part of one's own narrative, inevitably feels like an act of identity theft, even if the subject is a public official and the tone is generally sympathetic. Indeed, in *Gunslinger 1 & 2*, the poet Edward Dorn imagined the character of the Gunslinger, a philosophically minded vigilante who injures people simply by describing them. "It is dangerous to be named and makes you mortal," the Gunslinger declares. "If you have a name / you can be sold / you can be told / by that name leave, or come / you become, in short / a reference."[11]

To the degree that cyberspace makes it hard for us to escape from the electronic records of our missteps in the past, it increases our susceptibility to being described out of context in the future. The architecture of cyberspace, of course, might be redesigned to combat this danger. In the prologue, I mentioned the case of James Rutt, the Internet executive who used a special "scribble" feature offered by his chat group to delete a decade of his own postings. Like self-deleting e-mail, self-deleting chat technology should in time become more widely available. But as citizens search for ways of covering their electronic tracks, they will also struggle to reconstruct zones of privacy in cyberspace along a spatial as well as a temporal dimension. By blurring the boundaries between home and work, so that virtual space no longer corresponds to geographic space, cyberspace has altered the traditional private areas in which we can retreat from the observation and expectations of our employers and colleagues.

Remember Erving Goffman's distinction between the formal front region and the relaxed backstage, where individuals can

relieve the tensions that are an inevitable part of public performance. In workplaces where no backstage behavior is permitted, the strain on workers may be palpable: Goffman notes the
example of the American gas station, where customers feel free
to "define the whole station as a kind of open city for males, a
place where an individual runs the risk of getting his clothes
dirty and therefore has the right to demand full backstage privileges."[12] Denied a private space where they can work behind
closed doors, gas station mechanics are forced to perform under
constant scrutiny, which increases the tension of their job.

The flexible interplay between openness and privacy, between
the backstage and the front region, is preserved in successful
workplaces, from department stores and restaurants to law firms:
in order to function properly, all of us need a place to blow off
steam and to collect our thoughts. This is why tell-all memoirs,
which depict the president losing his temper and yelling at his
aides, must be considered a violation of privacy. Even though
presidents are not entitled to privacy in any ordinary sense,
they may, as Alan Westin observes, explode in temper tantrums
under the strains of office, and lash out with angry words that
they don't really mean. But the president's privacy in these un-
guarded moments is supposed to be respected, rather than betrayed, because everyone understands that these outbursts "make
possible the more measured speech of public presentation."[13]

Cyberspace, on balance, has reduced the backstage space
available to certain classes of American workers. To the extent
that the Internet makes it possible for those in writing-based
professions to work from home, it has created new opportunities for isolation from the observation of colleagues and employers: informal dress (or undress), sloppy sitting, mumbling,
and other self-involved behavior such as dancing around the

living room is perfectly permissible at home. Yet by blurring the boundaries between home and work, and by requiring workers to be accessible to their employers and colleagues by e-mail, cell phone, and beeper at all hours of the day and night, the new information technologies have also decreased our ability to control the conditions under which we are available to others.

Goffman's backstage is a social rather than a solipsistic place, designed not for solitary labor but for relaxed interactions with peers, free from the observation of superiors. And to the degree that all communications in cyberspace are monitorable and traceable by employers, cyberspace itself is now being treated as a front region rather than a backstage area, and its growth has decreased the private spaces in which we can express ourselves free from corporate scrutiny. The "profanity, open sexual remarks, elaborate griping," to use Goffman's examples, that used to be expressed around the watercooler or coffee machine are now, when recorded in e-mail, transformed from opportunities for letting off steam to potential evidence of illegal harassment. (As we saw in chapter 2, even griping and profanity e-mailed from home is now considered inappropriate conduct in the workplace.) There is nothing about the architecture of cyberspace itself that requires it to be treated as a front region rather than backstage area: if the legal definition of illegal harassment were refined, employers would be freer to draw more sensible distinctions between formal and informal speech, and to carve out areas of cyberspace—bulletin boards, chat rooms, and a range of private e-mails—as virtual backstage spaces, where workers could feel free to let down their guard. But in a world where communicating in cyberspace is increasingly unavoidable, and all communications in cyberspace are treated as public rather than private, workers will have fewer opportunities to collect themselves and put down their masks.

If recent changes in law and technology have increased privacy in some respects and decreased it in others, is there, on balance, any cause for alarm? Not everyone is convinced by the social value of privacy. Richard Posner, for example, argues that privacy can be inefficient, and can contribute to fraud and misrepresentation, because it allows people to conceal true but embarrassing information about themselves from other people in order to gain unfair social or economic advantage.[14] Richard Wasserstrom suggests that our insistence on leading dual lives— one public, the other private—can amount to a kind of deception and hypocrisy, and if we were less embarrassed by sexual and other private activities that have traditionally been associated with shame, we would have less to fear from disclosures of the self because we would have nothing that requires hiding.[15] In a more popular version of this argument, David Brin in *The Transparent Society* argues that the real threat to American liberty is not privacy but access to personal information. Rather than trying in vain to resist a world where ubiquitous video cameras mean that our lives are increasingly subject to public scrutiny, Brin argues, privacy advocates should focus instead on ensuring that all citizens have access to one another's videotapes. Brin quotes John Perry Barlow, the former Greatful Dead lyricist, and now an advocate of privacy in cyberspace, on behalf of the thesis that transparency might be preferable to privacy: "I have no secrets myself, and I think that everybody would be a lot happier and safer if they just let everything be known. Then nobody could use anything against them. But this is not the social norm at the moment."[16]

The defenders of transparency are confusing secrecy with privacy, even though secrecy is only a small dimension of privacy. ("Are you talking about something secret?" a colleague asked when she walked in on her teenage daughter's phone conversation. "No," the daughter replied. "It's private.") Even those who claim that society would be better off if people were less embarrassed about discussing their sexual activities in public still manage to feel annoyed and invaded when they are solicited by telemarketers during dinner. Moreover, the defenders of transparency have adopted a view of human personality as essentially unitary and integrated, so that social masks are viewed as a way of misrepresenting the true self. But Walter Mischel's work on character, which I discussed in chapter 4, suggests that this view of personality is simplistic and misleading, and that instead of behaving in a way that is consistent with a single character, people display different characters in different contexts.[17] I may (and do) wear different public masks when interacting with my students, my close friends, my family, and my dry cleaner. Far from being inauthentic, each of these masks helps me try to behave in a manner that is appropriate to the different roles demanded by these different social settings. If these masks were to be violently torn away, as Clarence Thomas's ordeal shows, what would be exposed is not my true self but the spectacle of a wounded and defenseless man.

If this "dramaturgical" view of character is correct, and if privacy is defined broadly, as the ability to protect ourselves from being judged out of context, then there are clearly political, social, and personal costs to the changes in the architecture of privacy that I have discussed. First, let's review the political costs. Privacy, I've argued throughout this book, makes public as well as social life possible: we can think each other disgusting, or wrongheaded, or depraved, and live together without resolving our differences. The political philosopher Judith Shklar gave a helpful example of the political value of privacy: in a democracy, Shklar argues, we don't need to know someone's title to avoid giving offense. The democratic honorific, Mr. or Ms., suggests that all citizens are entitled to equal respect, without revealing their rank or family background or professional accomplishments. Democracy is a space where citizens and strangers can interact without putting all their cards on the table; and privacy, as I suggested in chapter 4, allows citizens who disagree profoundly to debate matters of common concern without confronting their irreconcilable differences.

The historical movement toward less privacy for public officials, and more privacy for individual citizens, is a central achievement of a liberal democracy; but a properly functioning pluralistic society maintains a balance between privacy and publicity. "An effectively working pluralistic system will feel no need for complete publicity," Edward Shils argues in his study of privacy violations during the McCarthy era. "The mutual confidence of the elites of the different spheres of the corporate bodies within the spheres renders unnecessary the perpetual disclosure of the private affairs of rivals and fellow citizens."[18] Shils notes that secrecy is often the enemy of privacy, because "in order for secrets to be safeguarded, privacy must be invaded."[19] The effort to investigate potential breaches of security in the McCarthy era, Shils observes, inevitably led to serious violations of the privacy of innocent people: in the absence of direct evidence of wrongdoing, it was often the character and beliefs of suspected spies, rather than their actions, that were on trial. Every aspect of a suspect's life—his intimate friendships, casual acquaintances, loyalties, and affiliations—became relevant to the unbounded investigations. In investigating the dignitary offenses alleged by Anita Hill, Paula Jones, and less famous privacy victims, the accused harasser's thoughts, character, and private life were put on trial in similarly violating ways.

During periods of turmoil, when the culture of scandal and investigation comes to dominate politics, political rivals from each of the branches of government may try to expose each other's most intimate secrets. The imbalance that occurs when one of the three branches fails to respect the deliberative privacy of the other two is a consequence of what Benjamin Ginsberg and Martin Shefter call "a major new technique of political combat—revelation, investigation and prosecution."[20] In a world in which politics is criminalized, and political battles are transformed into legal battles, the institutional norms that ordinarily persuade the branches to give each other a measure of reciprocal privacy break down. The result—intense polarization, political extremism, ruined reputations, and a breakdown of public deliberation—are by now familiar. Because the prosecutorial culture creates episodic rather than permanent national scandals—as the McCarthy, Thomas, and Clinton dramas show—it is unlikely to bring government to a halt. But candid deliberations in Congress, the White House, and the judicial branch may be chilled by the overly intrusive surveillance of the other branches: at the height of the Lewinsky scandal, one White House aide described to me a "cone of totalitarianism" in the White House, where trusted friends would break off their conversations in midsentence for fear of receiving a subpoena.

There are social costs related to the erosion of privacy. The increased surveillance and monitoring that government officials experience in the political sphere are increasingly common in private workplaces as well, with similarly inhibiting effects on creativity and even productivity. Several surveys of the health effects of monitoring in the workplace have suggested that electronically monitored workers experience higher levels of depression, tension, and anxiety, and lower levels of productivity, than those who are not monitored.[21] It makes intuitive sense that people behave differently when they fear that their conversations may be monitored. As Stanley Benn has noted, the knowledge that you are being observed changes your consciousness of yourself and your surroundings: even if the topic of the conversation is not inherently private, your opinions and actions suddenly become candidates for a third party's approval or contempt.[22] Unsure about when, precisely, electronic monitoring may take place, employees will be more guarded and less spontaneous, and the increased formality of conversation and e-mail can make communication less efficient. (Indeed, language in primitive societies can be more formal, ceremonial, and elaborately polite than that of modern Americans, because people in primitive societies often have less privacy, and must take into account the effect of their words on third parties.[23]) Moreover, people in certain occupations tend to exaggerate the risks of public exposure: how many ambitious lawyers and law professors have changed their e-mailing habits in anticipation of Senate confirmation hearings that may never materialize?

In his preface to "Panopticon," Jeremy Bentham imagined the educational benefits of a ring-shaped "inspection-house," in which prisoners, students, orphans, or paupers could be subjected to constant surveillance. In the center of the courtyard was an inspection tower with windows facing the inner side of the ring. Supervisors in the central tower could observe every movement of the inhabitants of the cells, who were illuminated by natural lighting, but venetian blinds ensured that the supervisors could not be seen by the inhabitants. Michel Foucault described the purpose of the Panopticon in this way: "to induce in the inmate a state of conscious and permanent visibility that assures the automatic functioning of power."[24] Foucault argued that this condition of visible, unverifiable power, in which individuals have internalized the idea that they may always be under surveillance, is a defining characteristic of the modern age. It is not obvious that citizens in secular democracies at the beginning of the twenty-first century in fact perceive themselves to be under more intrusive surveillance than citizens, say, in the Middle Ages did: in an age of faith, there was no escape from the gaze of an unblinking God, whom the faithful conceived as the "reader of our entrails." Nevertheless, the modern phenomenon of Pantopticism, in which surveillance is technological rather than spiritual, reminds us that citizens work and interact very differently in circumstances where they are unsure about whether or not they are being watched.

Workers also experience a dignitary injury when they are treated like the inhabitants of the Panopticon, or like the dehumanized hero of *The Truman Show,* a character who has been placed on an elaborate stage set without his knowledge or consent, and whose every move, as he interacts with the actors who have been hired to impersonate his friends and family, is recorded by hidden video cameras. Spying on people covertly is an indignity, much like the indignity of harassment itself. It fails to treat its objects as fully deserving of respect, and instead treats them like animals in a zoo, deceiving them about the nature of their own surroundings. In this sense, privacy makes professional interactions possible: if workers come to feel they are being treated like children, it may be harder for them to maintain their professional roles as adults. This is why unwanted gazes can be professionally wounding: they convey the message that the person in question is viewed as a sex object rather than a professional. It is no coincidence that the intrusion on seclusion tort has been successfully invoked as a remedy both for covert electronic monitoring and for unwanted sexual advances in the workplace. In both cases, the dignitary injury is similar.

Finally, there are the personal costs of the erosion of privacy. Privacy is important not only, or even primarily, to protect individual autonomy but also to allow individuals to form intimate relationships. Without a commitment to privacy, the Harvard legal philosopher Charles Fried has written in one of the most thoughtful essays on the subject, "respect, love, friendship and trust" are "simply inconceivable."[25] Friendship and romantic love can't be achieved without intimacy, and intimacy, in turn, depends upon the selective and voluntary disclosure of personal information that we don't share with everyone else. Behind the protective shield of privacy, two individuals can relax the boundaries of self and lose themselves in each other. Because of their mutual self-exposure, friends and lovers are uniquely vulnerable to each other, which is why a serious fight in a friendship or a marriage can quickly escalate to a nuclear exchange. There is no greater threat to privacy than the trauma of being a party in a divorce suit or a child custody battle, during which all of one's most intimate secrets are exposed to public view. In her short story "The Other Two," Edith Wharton coolly describes a twice-divorced woman who finds herself serving tea to all three of her husbands at the same time. "She was 'as easy as an old shoe'—a shoe that too many feet had worn," Wharton writes. "Her elasticity was the result of tension in too many different directions. Alice Haskett—Alice Varick—Alice Waythorn—she had been each in turn, and had left hanging to each name a little of her privacy, a little of her personality, a little of the inmost self where the unknown god abides."[26]

Properly shielded, friendships and loving relationships provide us with opportunities for sharing confidences and testing ideas because we trust that our confidences won't be betrayed. (A friend is a person with whom I may be sincere," said Emerson. "Before him, I may think aloud."[27]) The decision to move from the private to the public sphere requires testing our intuitions against the responses of our most trusted friends and colleagues. To the degree that jokes, rough drafts, and written confidences can be exposed to public scrutiny, it is less likely that these confidences will be shared in the first place. Friendship, of course, will survive the new technologies of monitoring and surveillance: if I am afraid that my e-mail to my friends may be misinterpreted, I will take care to talk to my friends over the telephone or in person. But rather than behaving like citizens in totalitarian societies, and passively adjusting our behavior to the specter of surveillance, we should think more creatively about ways of preserving private spaces and sanctuaries in which intimate relationships can flourish.

Privacy is necessary, I've argued, to protect important social relationships—to make it possible for people to interact as citizens in the public square, as professionals in the workplace, and as friends, lovers, and family members in intimate group settings. But there is also an important case for privacy that has to do with the development of human individuality. "Without privacy there is no individuality," Leontine Young has noted. "There are only types. Who can know what he thinks and feels if he never has the opportunity to be alone with his thoughts and feelings?"[28] Any writer will understand the importance of reflective solitude in refining arguments and making unexpected connections: in an odd but widely shared experience, many of us seem to have our best ideas when we are in the shower. Indeed, studies of creativity show that it's during periods of daydreaming and seclusion that the most creative thought takes place, as individuals allow ideas and impressions to run freely though their minds in a process that can be impeded by the presence of others.[29] Virginia Woolf, in *A Room of One's Own,* insisted that "five hundred a year and a room with a lock on the door"[30] were necessary to write great fiction and poetry. The lack of money and privacy, Woolf declared, had made it hard for women writers to flourish for most of recorded history.

To develop their creative potential, individuals need refuge not merely from their families and colleagues but also from the overwhelming pressures toward social conformity. Mill, for example, worried that in the Victorian age the tyranny of social opinion was so powerful that "everyone lives under the eye of a hostile and dreaded censorship." "Society has now fairly got the better of individuality," Mill wrote, because "it does not occur" to the individual or the family "to have any inclination except for what is customary. . . . [T]hey exercise choice only among things commonly done; peculiarity of taste, eccentricity of conduct are shunned equally with crimes, until by dint of not following their own nature they have no nature to follow." Mill's bitterness at the tyranny of public opinion was influenced by the gossip and social stigma that he experienced as a result of his unconventional twenty-year friendship and intellectual partnership with Harriet Taylor, the wife of a friend, whom he eventually married and whom he hailed as his collaborator and soul mate in the dedication of his greatest book. The consequence of overly intrusive social regulation of private life, Mill recognized, is to deprive the individuals in question of the free development of their own tastes and inclinations, and to deprive society of the creative accomplishments that can be produced only under conditions of privacy.

I have tried to defend a classically liberal vision of privacy, finally, out of a belief that the limited government on which liberalism insists is the best way of respecting the dignity and equality of the individual. The original meaning of the word "person," as the sociologist Marcel Mauss notes in his essay on the evolving notion of personhood, was exclusively that of a mask. In Pueblo Indian cultures, Mauss observes, the symbol of personhood was the mask that members of the clan wore in sacred dramas. Etymologists of Latin explain the word *persona* as the mask through which (*per*) resounds (*sonare*) the voice of the actor; and Roman citizens were defined by the legal privileges that attached to the masks, names, and images of their ancestors.[31] In a society in which the representation of the self is equated with the self, stealing someone's iconic image can be tantamount to stealing his identity; this is why, in certain indigenous societies, people resist having their photographs taken. But in the Christian era, personhood was imbued with the idea of an interior soul, which became transformed by Enlightenment philosophers into the notion of the autonomous and individual self, defined by its own subjective consciousness. Mauss rather sweepingly summarizes the historical evolution of personhood in this way: "From a simple masquerade to the mask, from a 'role' (*personnage*) to a 'person' (*personne*), to a name, to an individual; from the latter to a being possessing metaphysical and moral value; from a moral consciousness to a sacred being; from the latter to a fundamental form of thought and action—the course is accomplished."[32]

By insisting that there are personal boundaries that the state may not overstep, interior regions into which it cannot penetrate, liberalism expresses its respect for the inherent dignity, equality, individuality, interiority, and subjectivity of the individuals who compose it. Inviolability is a form of equality; people who are less than equal are people who can be violated. A liberal state respects the fact that each individual has some precious and incommensurable interior essence that must be protected from official scrutiny. But when a state goes beyond the negative role of refraining from violating the dignity and privacy of its citizens, and uses its power to punish citizens for violating one another's dignity and privacy, it may encourage a degree of surveillance and exposure far greater than the indignities it seeks to avoid.

For this reason, I have tried throughout this book to make a case for the superiority of norms over law in protecting privacy, except in extreme cases. If I betray my friends' confidences, I won't have very many friends, and if I behave inappropriately in the workplace, gossip will usually ensure that everyone knows about my behavior. Given the complexity of appropriate responses to inappropriate speech and conduct, employers should be freer than they are now to choose among a range of responses, from talking quietly to the offender to applying formal discipline. Indeed, in a pluralistic society, different employers in different occupations should have the option of protecting privacy in different ways: the military should be free to monitor e-mail and forbid fraternization; universities might regulate conduct between professors and students but refuse to regulate speech under any circumstances; and different corporations could strike different balances, based on the nature of their product and the kind of employees they hope to attract. This flexibility, in which privacy is enforced primarily through social disapproval, backed by voluntary choices made by employers, is something that the current legal regime refuses to allow.

Thanks to the welcome change in social norms that sexual harassment law helped to precipitate, legal regulations are less important in maintaining civility at work today than they were a generation ago. In the wake of Anita Hill and Paula Jones, men and women now realize that boorish behavior in the workplace is likely to have disastrous consequences in terms of public embarrassment; and if the excesses of harassment law have restored a Victorian caution to the way that men and women interact at work, this is hardly a bad thing. The question of when, precisely, norms shifted dramatically enough to permit a relaxation of legal prohibitions is open to debate, but it is now clear that collateral costs of legal regulation are too great, and the boundaries of privacy too delicate, for indignities that have no tangible employment consequences to be punished primarily through law.

To those who fear that social disapproval may not be adequate to enforce norms of sexual propriety if legal regulations were relaxed, history should provide some comfort. Soon after President Clinton was acquitted by the Senate with strong public support, he was booed at a soccer game. And consider the trial of Queen Caroline of England in 1820, the most famous political sex trial of the nineteenth century, which suggests that when prosecutors try to prove adultery by subpoenaing bodyguards and confidential aides, the public may prefer social to legal punishment.

The arranged marriage between the future George IV and Princess Caroline of Brunswick was doomed from the start. "Harris, I am not well, pray get me a glass of brandy," the Prince observed upon being introduced to his bride. A few days after the wedding, the couple managed to conceive a daughter; George then wrote his wife a letter demanding that they set up separate households. Absolved of her conjugal duties, Caroline left the country for a five-year tour of the Continent and the Middle East. While in Milan, she appointed as her "courier" an Italian manservant named Bartolommeo Pergami, and rumors of their debaucheries soon made their way back to London. When Caroline returned to England after the death of George III, the new king asked Parliament to investigate his wife, and the prime minister proposed a bill to impeach Caroline for "a licentious, disgraceful and adulterous intercourse," dissolving her marriage and depriving her of her title.

During the hearings that followed in the House of Lords, so many of the queen's bodyguards, chambermaids, and traveling companions were ordered to testify that her lawyer charged that there was a vast Italian "conspiracy" against her. A confidential servant named Theodore Majocchi reported that on a sailing ship from Jaffa to Italy, the princess had slept with Pergami in a shipboard tent, and that he had heard "the creaking of a bench." A royal chambermaid, Louisa Demont, said that she had seen semen-stained sheets.

The queen's lawyers made much of the prosecution's refusal to call the lady-in-waiting who had allegedly made the incriminating bed. In his summation, Thomas Denmont, the queen's counsel, thundered: "Did their lordships suppose that those agents who had collected together a set of her Majesty's discarded servants, who had ransacked filthy clothes-bags, who had raked into every sewer, pried into every water-closet, who attempted to destroy all the secrecies of private life . . . did their lordships imagine that they who had resorted to such mean and filthy practices would have stopped short of producing" the witness who could "prove the marks?"

At the end of the trial, a slim majority of the House of Lords voted for the bill to impeach the queen. Nevertheless, the Lords decided to shelve the bill at the last minute, influenced by the demonstrations of an angry London mob. Popular sentiment favored the queen, notes Caroline's biographer, Roger Fulford, not out of a belief in her innocence but because of "a revolution against authority."[33]

After Queen Caroline's prosecution was abandoned, delirious Londoners cheered her in the streets. But social disapproval has a way of catching up with private misconduct in the end. Eight months later, when Caroline's husband was formally crowned king and she attempted to force her way into Westminster Abbey, the mob blocked her entrance with cries of "Shame! Shame!" The queen had confused the crowd's dislike of her tormentors with affection for her.

—

In arguing that law can sometimes do more harm than good when it tries to remedy invasions of privacy, it may seem that I have written about two different subjects: on the one hand, the misdefinition of certain kinds of sexual harassment as sex discrimination rather than invasions of privacy; on the other, the recent incursions, by law and technology, into our ability to decide how much of ourselves to reveal to others. But in fact both developments are illustrations of the same subject, and that is the importance of maintaining private spaces to protect individuals from being judged out of context in a world of fleeting attention spans. The expectation that men and women should treat one another with respect at work insists that all Americans are entitled to be regarded as self-defining individuals rather than as prisoners of sexual stereotypes and generalizations. The ideal of privacy, similarly, insists that individuals should be allowed to define themselves, and to decide how much of themselves to reveal or to conceal in different situations. Except, perhaps, on a desert island, the notion that any individual can exercise complete control over the conditions of his or her self-definition is, in practice, unattainable. In society, it is impossible not to have your dignity assaulted by the unwanted gazes of others, and not to be misdefined and misjudged and wrenched out of context. Society is an orgy of judgments and misjudgments. But by respecting the boundaries between public and private speech and conduct, a liberal state can provide sanctuaries from the invasions of privacy that are inevitable in social interactions. And as technology allows us to reconstruct private spaces at home, at work, and in cyberspace, law should not stand in its way.

The invasions of privacy I have discussed in this book are part of a larger crisis in America involving the risk of mistaking information for knowledge in a culture of exposure. We are trained in this country to think of all concealment as a form of hypocrisy. But we are beginning to learn how much may be lost in a culture of transparency: the capacity for creativity and eccentricity, for the development of self and soul, for understanding, friendship, and even love. There are dangers to pathological lying, but there are also dangers to pathological truth-telling.

Privacy is a form of opacity, and opacity has its values. We need more shades and more blinds and more virtual curtains. Someday, perhaps, we will look back with nostalgia on a society that still believed opacity was possible and was shocked to discover what happens when it is not.

Afterword to the Vintage Edition

Since *The Unwanted Gaze* was published, the protection of privacy has emerged as one of the most salient political, cultural, and technological issues around the globe. As our lives are increasingly lived in cyberspace, citizens are beginning to experience the misunderstandings and indignities that occur when personal information is taken out of context. In particular, we are learning that privacy is threatened far more when information is collected in personally identifiable ways than when it is not.

This is lesson of DoubleClick, Inc., which was forced to retreat from its plan to compile profiles linking individuals' names and addresses to detailed records of their online and offline purchases. And other companies that have attempted to collect and store personally identifiable data have faced similar opposition.

For example, Globally Unique Identifiers, or GUIDs, are making it possible to link every document we create, every message we e-mail and every chat we post with our real-world identities. GUIDs are a kind of serial number that can be linked with our name and e-mail address when we register online for a product or service. Not long ago, RealJukebox, one of the most popular Internet music players, with 30 million registered users, became a focus of media attention when privacy advocates noted that the player could relay information to its parent company, RealNetworks, about the music each user downloaded, and that this could be matched with a unique ID number that pinpointed the user's identity. The company insisted that it had never, in fact, matched the GUIDs with the data about music preferences. Nevertheless, hours after the media outcry began, RealNetworks disabled the GUIDs to avoid a DoubleClick-like public relations debacle. Similarly, as soon as consumers complained that Sprint's wireless Web phone was revealing individual telephone numbers to pre-programmed Web sites, Sprint promptly disabled the feature.

These examples suggest that political and economic pressure can stop especially intrusive forms of data collection in their tracks. And they are signs of a healthy dialogue between consumers, privacy advocates, and dot-com investors.

Politics, of course, is not the only response to the invasions of privacy I discuss in the book: there is a vigorous legal debate as well. Privacy advocates are now focusing their energies on requiring a higher standard of consent for the collection of personally identifiable information than for other information. I'm more optimistic, however, about using law to protect individuals from intrusive forms of state surveillance. When the FBI created a search engine called Carnivore, which allows the government to scan all of the e-mail traffic on an Internet Service Provider in search of particular messages that have been identified by a court order, conservative and liberal Senators and Representatives joined in expressing concern. If Carnivore operates as the FBI promises, it might be considered the paradigm case of a reasonable search, since it reveals only the information it is authorized to reveal and nothing more. But in the hands of rogue agents, Carnivore could operate very much like the general warrants that the framers of the Fourth Amendment meant to forbid.

In addition to political and legal responses, there is growing interest in technological self help. At the time this book was originally published, privacy enhancing technologies were relatively esoteric—used by only a handful of especially zealous privacy afficionados, such as my former student, K., who wears black boots and green army fatigues and spends much of the day covering his electronic tracks. Aware that files and e-mail can be resurrected from his hard drive even after they are ostensibly deleted, K. uses a suite of security tools called Kremlin. Every time K. turns off his computer, Kremlin does a "secure total wipe" of his 20 gigabyte hard drive, scribbling electronic graffiti, in the form of zeroes and ones, over all the free space so that any lurking, partly deleted files will be rendered illegible. This takes more than an hour. K. also uses Kremlin to encrypt his personal documents in a secure folder on his hard drive, and he carefully chose a nonsense password, garbled with upper- and lowercase letters and numbers, so that it can't easily be cracked by a "brute force attack program" that might hypothetically bombard his computer with millions of random words generated from an electronic dictionary. But privacy enhancing technologies are becoming more user friendly; and it's not hard to imagine that in a decade or so, citizens will be able to choose from a broad menu of options, at home and work. All this will make it easier for individual citizens to protect themselves against the dangers of being judged out of context by taking steps to cover their own tracks.

Perhaps the most important task for defenders of privacy over the next decades will be to continue to make a case for why privacy matters. In the book, I stress the importance of not being judged out of context, but protection against misjudgment is not the only purpose of privacy. In an illuminating response to the book, published in a review symposium in *The Georgetown Law Journal*, Robert Post of the University of California at Berkeley has identified three concepts of privacy in *The Unwanted Gaze*: the first connects privacy to dignity, the second to autonomy, and the third to the creation of knowledge. Post suggests that only the first, privacy as dignity, is distinctively a problem of privacy. The second, he says, is best conceived as a form of liberal freedom from totalizing state surveillance; while the third conception, the problem of misrepresentation or incomplete understanding, shouldn't be understood as a problem of privacy at all, because people can be misrepresented or judged out of context on the basis of public or private information.

Privacy conceived as a form of dignity, as Post suggests, focuses on the "social forms of respect that we owe each other as members of a common community"; while privacy as autonomy concerns the individual's ability to maintain a sphere of immunity from social norms and regulations. Although I agree that privacy is necessary to protect dignity and autonomy, I'd like to defend the thesis that the problem of being judged out of context is distinctively a problem of privacy. When private information is taken out of context, the social judgments that result are, in fact, more damaging to the individual, and more likely to lead to cognitive errors on the part of society, than the social judgments that result when public information is taken out of context. When private information is taken out context, the only way to try to put the information in a broader context is to reveal more private information, which only increases the risk of misinterpretation, because certain kinds of private information can only be understood in a context of intimacy.

Let's think, first of all, about misrepresentations on the basis of public information that is taken out of context. This is the football player whom everyone remembers for making a single bone-headed play; or the author who receives an unfair review of a deserving book. It's true that being misrepresented is distressing, whether or not it is caused by the revelation of private facts. But if I'm a football player who makes a boneheaded play, I can improve my image by playing better next time. If I'm an author who gets a bad review, I can point to a better one. If I'm a president who gets sick at a state dinner, I can win a gulf war. Misjudgment on the basis of private information requires the involuntary disclosure of more private information, while misjudgment on the basis of public information is more easily countered by behaving in accordance with my public role.

Now let's think about misrepresentation on the basis of private information that has been taken out of context. In this situation, both the injury and the remedy look very different. Misjudgment on the basis of private information is more likely to damage the individual than misjudgment on the basis of public information. In the former case, the cure is worse than the disease, and the misjudgment can only be corrected by the revelation of more private information, which leads to further misunderstanding and further harm to dignity and autonomy.

Consider the case of Lawrence Lessig's e-mail. Lessig said he didn't consider the publication of the e-mail itself to be especially invasive, but in order to put his joke in context he—and I—had to reveal the fact that he had been listening to the singer Jill Sobule. But the revelation of this contextual backstage information distorted the information market in new ways. Readers of *The Unwanted Gaze* now know that Lessig wasn't biased against Microsoft, but they think of him as the kind of person who listens to Jill Sobule. He isn't just that kind of person, however: he also has a passion for Gregorian chant. And now that I've revealed that my friend Lessig likes to listen to Gregorian chant, perhaps I've misrepresented him further: after all, he isn't just a Jill Sobule person and a plainsong person. The point isn't that Lessig's music preferences themselves are in any way embarrassing; the point is that once the backstage curtain is lifted, Lessig and those who know him can only put the information in context by revealing even more private information.

But the world isn't entitled to know about Lessig's music preferences, not merely because the world has no time to understand Lessig in all of his wondrous dimensions—this is the attention span problem—but because Lessig shouldn't have to justify his music preferences to the world. Knowledge must be earned by the slow, reciprocal sharing of personal information, which leads, in turn, to greater intimacy, understanding and trust. This is the process that is short-circuited when private information is taken out of context. To understand Lessig's joke required more attention than the public was able to give, and more intimacy than the public was entitled to demand.

Here, then, are the range of injuries that result from being observed out of context in spaces that should be considered private. To the extent that the observation results from unreasonable state extraction of intimate personal information—the subpoenas that exposed the e-mails of Lessig and Lewinsky—it is an offense against liberal freedom in the traditional sense of threatening the boundaries between the individual and the state. When the state engages in low-level pervasive surveillance—such as the cameras that were recently deployed at the Superbowl to survey fans as they entered the stadium and match their faces with data banks of suspected criminals—it threatens autonomy in a similar way. Surveillance by all-seeing advertising networks such as DoubleClick can also threaten autonomy, insofar as it inhibits our ability to express our-selves freely in spaces that should be considered private.

To the degree that observation results from the betrayal of a friend or former lover, such as Joyce Maynard's tell-all memoir about J.D. Salinger, it represents a breach of confidence; and an offense against socially constructed norms of dignity, like the one I would experience if a friend snapped a photo of me at a nude beach. If social norms changed and tell-all memoirs (or public nudity) became more common than they are now, such a betrayal might no longer represent an offense against dignity, but it would still represent an offense against autonomy. If individuals can't form relationships of trust without fear that their confidences will be betrayed, the uncertainty about whether or not their most intimate moments are being recorded and monitored will make intimacy impossible; and without intimacy, there will be no oppor-tunity to develop the autonomous, inner-directed self that defies social expectations rather than conforming to them.

I'm gratified, finally, by the reaction of some of the privacy vic-tims whose stories are told in the book. Monica Lewinsky is one of the heroes of *The Unwanted Gaze,* and I was pleased that she agreed with the thesis that privacy protects us from being judged out of context in a world of short attention spans, a world in which information can be confused with knowledge. "The world knew that I bought *Vox,*" she told me, "but they didn't know that I'd also recently read a nineteenth-century set of the work of Charles Dick-ens." The Lewinsky affair is now part of history, but for revealing the vulnerabilities that all of us face in a world where our intimate reading, writing and gossip in cyberspace can be recorded and monitored by strangers, we have Monica Lewinsky to thank.

March 2001
Washington, D.C.

Acknowledgments

This book would not have been written without the encouragement of Jonathan Karp at Random House, who suggested that I write about privacy, waited patiently for the manuscript to materialize, and offered enthusiasm and shrewd advice from beginning to end. At *The New Republic,* Martin Peretz gave me my first job after I graduated from law school and has provided a home for nearly a decade, characterized by complete freedom and reciprocated loyalty. Invaluable conversations with Leon Wieseltier helped to refine the argument of the book and pointed me toward the title. At *The New Yorker,* Tina Brown and Dorothy Wickenden generously encouraged "Is Nothing Private?," parts of which were incorporated into chapter 1. David Remnick and Lee Aitkin provided valued support for "Jurisprurience," which became a launching pad for chapter 4.

I'm grateful to my colleagues at the George Washington University Law School for including me in an ideal community for teaching and writing and for vigorous criticisms of chapter 3 that emerged at a faculty colloquium. An early draft of the manuscript was presented to the privacy colloquium of the 1999 Summer Humanities Institute at Dartmouth College, and conversations with Jeff Weintraub and his colleagues converged in the Epilogue.

Peter Berkowitz, Rosa Ehrenreich, Julie Hilden, Neal Katyal, Lawrence Lessig, Joanna Rosen, Marc Rotenberg, Ben Sherwood, Peter Swire, Eugene Volokh, and Ben Wittes read the manuscript with extraordinary care and improved the arguments immeasurably. Their generosity has helped me to understand the connections between friendship, privacy, and self-expression. Ruth Shalit has always been my best editor; her influence and sparkling intelligence are reflected on every page. Ingrid Abrash and Douglas Hudson were admirable research assistants. Christopher Reed of the George Washington University Law Library supplied books whenever they were needed. Devon Spurgeon provided boundless support and encouragement that made the writing process, for me at least, a pleasure. My agent, Amanda Urban, steered me toward Jon Karp and has been a formidable advocate.

Notes

PROLOGUE: THE UNWANTED GAZE

1. Andrew Morton, *Monica's Story* (New York: St. Martin's Press, 1999), p. 215.

2. Janet Kornblum, "Amazon.com Feature Fuels Privacy Fears," *USA Today,* August 26, 1999, p. 1A.

3. John Schwartz, "Could Online Chat Byte Back?: A CEO's Rush to Edit the Past Illustrates Perils of Posting," *The Washington Post,* October 7, 1999, p. A1.

4. Milan Kundera, *Testaments Betrayed* (New York: HarperCollins, 1995), pp. 260–61.

5. Erving Goffman, *The Presentation of Self in Everyday Life* (Garden City, N.Y.: Doubleday, 1959), p. 128.

6. For a fair-minded discussion, see Judith Wagner DeCew, *In Pursuit of Privacy: Law, Ethics, and the Rise of Technology* (Ithaca, N.Y.: Cornell University Press, 1997), p. 81 (Chapter 5, The Feminist Critique of Privacy).

7. Catharine A. MacKinnon, *Toward a Feminist Theory of the*

State (Cambridge, Mass.: Harvard University Press, 1989), p. 191.

8. Ibid., p. 194.

9. Morton, *Monica's Story,* p. 160.

10. See, e.g., Planned Parenthood v. Casey, 505 U.S. 833, 856 (1992). ("The ability of women to participate equally in the economic and social life of the Nation has been facilitated by their ability to control their reproductive lives.")

11. Erving Goffman, *Behavior in Public Places: Notes on the Social Organization of Gatherings* (New York: Free Press, 1963), p. 84.

12. Ibid., p. 85.

13. Ibid., p. 116 (quoting *The Laws of Etiquette,* by "A Gentleman," p. 62 [1836]).

14. Some Iranian women writers have defended the requirement that women veil their bodies and men avert their eyes as a way of protecting women from being objectified by the male gaze. See, e.g., Farzaneh Milani, *Veils and Words: The Emerging Voices of Iranian Women Writers* (Syracuse, N.Y.: Syracuse University Press, 1992), p. 8. ("The veil also protected women. It blocked the masculine gaze, subverted man's role as the surveyor, and removed women from the category of object-to-be-seen.") See also Alev Lytle Croutier, *Harem: The World Behind the Veil* (New York: Abbeville Press, 1989), p. 17. ("The veil is a way to secure personal liberty in a world that objectifies women. . . . *Hijab* [veiling] protects women from the male gaze and allows them to become autonomous subjects.")

15. See Indiana State University Sexual Harassment Policy, available at <http://web.indstate.edu/aaction/sex_har_pol.html>.

16. Goffman, *Behavior in Public Places,* pp. 137–38, quoting

Cornelia Otis Skinner, "Where to Look," in *Bottoms Up!*, pp. 29–30 (1955).

17. Menachem Elon, *Jewish Law: History, Sources, Principles: Ha-Mishpat Ha-Ivri* (Philadelphia: Jewish Publication Society, 1994), pp. 1859–60. "No one may place an entrance . . . opposite the entrance [of another] . . . nor a window opposite [another's] window," the Mishna states. "No one shall open up windows facing a jointly owned courtyard." Ibid.

18. Ibid., p. 1860.

19. *"Hezzek Re'iyyah,"* Encyclopedia Talmudit (Jerusalem, 1975), vol. 8, col. 673. This formulation of the principle is derived from discussions in Maimonides, Jacob ben Asher, and Solomon ben Adret. I am indebted to Leon Wieseltier for this translation, for the alternative that appears as the epigraph, and for walking me through the medieval sources.

20. Elon, *Jewish Law,* p. 943.

21. Anita L. Allen, *Uneasy Access: Privacy for Women in a Free Society* (Totowa, N.J.: Roman & Littlefield, 1988), p. 128.

22. See, e.g., Laura Mulvey, "Visual Pleasure and Narrative Cinema," in *Visual and Other Pleasures* (Bloomington: Indiana University Press, 1989), p. 190. ("In a world ordered by sexual imbalance, pleasure in looking has been split between active/male and passive/female. The determining male gaze projects its fantasy onto the female figure, which is styled accordingly.") See also E. Ann Kaplan, "Is the Gaze Male?," in *Powers of Desire: The Politics of Sexuality,* Ann Snitow et al., eds. (New York: Monthly Review Press, 1983), p. 309.

23. Goffman, *Behavior in Public Places,* pp. 134–35.

24. John Stuart Mill, *On Liberty* (Indianapolis: Bobbs-Merrill, 1956), p. 114. As Mill's critics have noted, the line between

self- and other-regarding conduct is not always precise or easy to draw, but the distinction remains a useful rule of thumb for limiting state power in a way that advances the cause of individual dignity. See, e.g., John Gray, *Mill on Liberty: A Defence* (London: Routledge, 1996), pp. 48–57.

CHAPTER 1: PRIVACY AT HOME

1. Morton, *Monica's Story,* p. 257.
2. R. Postgate, *That Devil Wilkes* (New York: Vanguard Press, 1929), pp. 171–78.
3. Eric Schnapper, *Unreasonable Searches and Seizures of Private Papers,* 71 VA. L. REV. 869, 890 n. 120 (quoting A LETTER TO THE RIGHT HONORABLE EARLS OF EGREMONT AND HALIFAX, HIS MAJESTY'S PRINCIPAL SECRETARIES OF STATE, ON THE SEIZURE OF PAPERS 25–31 (London, 1763) [*hereinafter* WILKES'S LETTER]).
4. See WILKES'S LETTER, *supra* note 3, at 19–21.
5. Schnapper, *supra* note 3, at 883 n. 87 (citing Wilkes v. Wood, Lofft 1, 1, 98 Eng. Rep. 489, 490 (C.P. 1763)).
6. Olmstead v. United States, 277 U.S. 438, 474 (1928) (Brandeis, J., dissenting).
7. Boyd v. United States, 116 U.S. 616, 631–32 (1886).
8. Huckle v. Money, 19 How. St. Tr. 1404, 1405, 95 Eng. Rep. 768 (C.P. 1763).
9. Warden v. Hayden, 387 U.S. 294, 322 (1967) (Douglas, J. dissenting).
10. David Brin, *The Transparent Society: Will Technology Force Us to Choose Between Privacy and Freedom?* (Reading, Mass.: Addison-Wesley, 1998), p. 69.
11. In a desperate effort to get the police to respond to an attack

on her son, the woman told a dispatcher he had been shot rather than beaten. See Weber v. Dell., 804 F.2d 796, 798–99 (2d Cir. 1986).

12. See, e.g., James X. Dempsy, *Communications Privacy in the Digital Age: Revitalizing the Federal Wiretap Laws to Enhance Privacy,* 8 ALB. L.J. SCI. & TECH. 65, 75 (1997).

13. Michael Adler, *Cyberspace, General Searches, and Digital Contraband: The Fourth Amendment and the Net-Wide Search,* 105 YALE L.J. 1093 (1996).

14. Lawrence Lessig, *Reading the Constitution in Cyberspace,* 45 EMORY L.J. 869, 882 (1996).

15. Arnold H. Loewy, *The Fourth Amendment as a Device for Protecting the Innocent,* 81 MICH. L. REV 1229 (1983).

16. Raphael Winick, *Searches and Seizures of Computers and Computer Data,* 8 HARV. J.L. & TECH. 75, 105–106 (1994) (citing United States v. Tamura, 694 F.2d 591, 595–96 (9th Cir. 1982)) (discussing the Intermingled Records Doctrine in the *Tamura* and *Steven Jackson Games* cases). See also United States v. Shilling, 826 F.2d 1365, 1369 (4th Cir. 1987).

17. Winick, *supra* note 16, at 107.

18. Oziel v. Superior Court of Los Angeles County, 273 Cal. Rptr. 196, 207 (Ct. App. 1990).

19. Hanlon v. Berger, 119 S. Ct. 1706 (1999).

20. Wilson v. Layne, 119 S. Ct. 1692 (1999).

21. RESTATEMENT (SECOND) OF TORTS § 652C (1977).

22. RESTATEMENT (SECOND) OF TORTS § 652E (1977).

23. See, e.g., Bailer v. Erie Ins. Exch., 687 A.2d 1375 (Md. Ct. App. 1997).

24. RESTATEMENT (SECOND) OF TORTS § 652D (1977).

25. Haynes v. Alfred A. Knopf, Inc., 8 F.3d 1222, 1232 (7th Cir. 1993).

26. Sipple v. Chronicle Publishing Co., 201 Cal. Rptr. 665, 670 (Cal. App. 1st. 1984).

27. Ferdinand David Schoeman, *Privacy and Social Freedom* (New York: Cambridge University Press, 1992), p. 155.

28. See Anita L. Allen, *Lying to Protect Privacy*, 44 VILL. L. REV. 161, 178–79 (1999).

29. See, e.g., Sanders v. Am. Broad. Co., 978 P.2d 67, 69 (Cal. 1999) (holding that "[i]n an office or other workplace to which the general public does not have unfettered access, employees may enjoy a limited, but legitimate, expectation that their conversations and other interactions will not be secretly video-taped by undercover television reporters").

30. Harry Kalven, Jr., *Privacy in Tort Law—Were Warren and Brandeis Wrong?*, 31 LAW & CONTEMP. PROBS. 326, 336 (1966). The coup de grâce occurred in 1975, when the Supreme Court held that the First Amendment allows the publication of any matters contained in public records, even if the publication would offend the sensibilities of a reasonable person. See Cox Broadcasting Corp. v. Cohn, 420 U.S. 469, 496 (1975).

31. Cason v. Baskin, 155 Fla. 198, 202 (1944).

32. Ibid., pp. 204–5.

33. Cason v. Baskin, 159 Fla. 31, 40 (1947).

34. Campbell v. Seabury Press, 486 F. Supp. 298, 301 (1979).

35. Anonsen v. Donahue, 857 S.W.2d 700 (Tex. App. Rt. 1993).

36. RESTATEMENT (SECOND) OF TORTS § 652D (1977).

37. Robert C. Post, *The Social Foundations of Privacy: Community and Self in the Common Law Tort*, 77 CALIF. L. REV. 957, 968 (1989).

38. Miller v. California, 413 U.S. 15, 24 (1972).

39. Michaels v. Internet Entertainment Group, Inc., 5 F. Supp.2d 823 (C.D. Cal. 1998). The court granted a preliminary injunction against IEG based on Michaels's likelihood of success on a copyright claim. Nevertheless, the court later granted summary judgment in favor of Paramount, after its tabloid news program, *Hard Copy,* had aired portions of the tape. The court concluded that the newsworthiness of the story protected Paramount from liability, and use of the plaintiff's name was part of the story and not a violation of the right to publicity. In the court's view, the First Amendment interest in a free press outweighed the right to privacy. See Michaels v. Internet Entertainment Group, Inc., 1998 U.S. Dist. LEXIS 20786, 48 U.S.P.Q.2d (BNA) 1891 (C.D. Cal. September 11, 1998).

CHAPTER 2: PRIVACY AT WORK

1. Andrew Morton, *Monica's Story,* p. 221.
2. See Elizabeth Weise, " 'Self-destruct' E-mail Offers Virtual Privacy," *USA Today,* October 7, 1999, p. 1A.
3. See Lawrence Lessig, *The Architecture of Privacy* (1998), available at <http://cyber.law.harvard.edu/lessig.html>.
4. Lisa Guernsey, "The Web: New Ticket to a Pink Slip," *The New York Times,* December 16, 1999, p. E1.
5. "The Surveillance Society," *The Economist,* May 1, 1999.
6. See, e.g., Scott A. Sundstrom, *You've Got Mail! (And The Government Knows It): Applying The Fourth Amendment to Workplace E-Mail Monitoring,* 73 N.Y.U.L. Rev. 2064, 2064–65 & n. 2 (1998) (citing "E-Mail Common in Workplace, But Usage Policies Lacking," *Newsbytes,* February 12, 1996, available in Westlaw, 1996 WL 7907264, discussing re-

sults of Society for Human Resource Management study stating that 7.7 percent of companies surveyed perform random employee e-mail reviews); Amitai Etzioni, "Some Privacy, Please, for E-Mail," *The New York Times*, November 23, 1997, p. C12 (stating that "various surveys" agree that over one third of employers monitor employees, generally through e-mail spot checks). In 1993, a study of 301 companies found that 22 percent of the employers searched their employees' "computer files, voice mail, electronic mail, or other networking communications." See Kevin J. Baum, *E-Mail in the Workplace and the Right of Privacy*, 42 VILL. L. REV. 1011, 1016–17 & n. 26 (1997) (citing Andrea Bernstein, "Who's Reading Your E-mail?," *Newsday*, July 15, 1996, p. A21).

7. See Sundstrom, *supra* note 6, at 2065 & n. 3 (citing Assentor Fact Sheet (visited September 7, 1998) <http://www.sra.com/industry_sectors/is_assentor.html>; Thomas Hoffman, "Brokers Can Monitor E-mail More Easily," *Computerworld*, July 20, 1998, p. 38; Carl S. Kaplan, "Big Brother as a Workplace Robot," *CyberTimes* (*The New York Times on the Web*) (July 24, 1997) <http://www.nytimes.com/library/cyber/law/072497law.html>).

8. Olmstead v. United States, 277 U.S. 438, 465 (1928). See, e.g., Lawrence Lessig, *Reading the Constitution in Cyberspace*, 45 EMORY L.J. 869, 872–73.

9. Katz v. United States, 389 U.S. 347 (1967) (Harlan, J., concurring).

10. Christopher Slobogin and Joseph E. Schumacher, *Reasonable Expectations of Privacy and Autonomy in Fourth Amendment Cases: An Empirical Look at Understandings Recognized and Permitted by Society*, 42 DUKE L.J. 727, 737 (1993).

11. Ibid., p. 738.

12. Ibid.

13. O'Connor v. Ortega, 480 U.S. 709, 725 (1987) (plurality opinion).

14. Slobogin and Schumacher, *supra* note 10, at 741–42.

15. See Skinner v. Railway Labor Executives' Assn., 489 U.S. 602, 634 (1989); National Treasury Employees Union v. Von Raab, 489 U.S. 656, 677 (1989), cited in Slobogin and Schumacher, *supra* note 10, at 742 n. 55.

16. Slobogin and Schumacher, *supra* note 10, at 742.

17. Albert W. Alschuler, *Interpersonal Privacy and the Fourth Amendment*, 4 N. ILL. L. REV. 1, 24 (1983).

18. Smith v. Maryland, 442 U.S. 735, 743–44 (1979) (emphasis removed).

19. Ibid. at 750. See also United States v. Miller, 425 U.S. 435, 455 (1976) (Marshall, J., dissenting) (citing California Bankers Assn. v. Shultz, 416 U.S. 21, 96 (1974) (Marshall, J., dissenting)).

20. See Alschuler, *supra* note 17, at 6–8 & n. 12. Alschuler stresses that when the government intrudes on property interests in persons, houses, papers, and effects, with or without physical trespass, judges shouldn't have to inquire into cultural expectations of privacy. They should speculate about cultural expectations, he argues, only when evaluating invasions of privacy that take place outside this property-based Fourth Amendment core.

21. See Zurcher v. Stanford Daily, 436 U.S. 547 (1978); 42 U.S.C. § 2000aa(a) (1994 & supp. 1998) (intending to overrule *Zurcher* by statute). Searches of the offices of authors and publishers are permitted only if there is probable cause to believe that the person possessing the materials is involved in the crime

to which the materials relate, or there is reason to believe that a seizure of the material is necessary to prevent death or serious bodily injury, or it seems likely that a subpoena would be defied, or would result in the loss of evidence.

22. Steven Jackson Games Inc. v. United States Secret Service, 816 F. Supp. 432, 440–41 (1993).

23. See O'Connor, 480 U.S. at 725–26.

24. Ibid., p. 712.

25. Ortega v. O'Connor, 146 F.3d 1149, 1161 (9th Cir. 1998) (second appeal after remand).

26. See O'Connor, 480 U.S. at 712, 728.

27. See, e.g., Ortega v. O'Connor, 764 F.2d 703, 705–06 & n. 1 (9th Cir. 1985), rev'd 480 U.S. 709 (1987).

28. See O'Connor, 480 U.S. at 713.

29. Ibid., pp. 717–20.

30. Ibid., pp. 724–29.

31. Ortega, 146 F.3d at 1163–64.

32. Whitfield Diffie and Susan Landau, Privacy on the Line: The Politics of Wiretapping and Encryption (Cambridge, Mass.: MIT Press, 1998), pp. 128–29.

33. United States v. Maxwell, 45 M.J. 406, 418 (1996).

34. Bohach v. Reno, 932 F. Supp. 1232, 1234 & n. 2 (1996) (quoting Laurie Thomas Lee, Watch Your E-Mail! Employee E-Mail Monitoring and Privacy Law in the Age of the "Electronic Sweatshop," 28 J. MARSHALL L. REV. 139, 148 (1994).

35. Bourke v. Nissan Motor Corp., No. YC003979, slip op. (D.C. Cal. 1993).

36. Baum, supra note 6, at 1039.

37. S. Elizabeth Wilborn, Revisiting the Public/Private Distinction: Employee Monitoring in the Workplace, 32 GA. L. REV. 825,

844–46 (1998) (The requirements for the intrusion on seclusion tort "typically defeat[] the employee's tort claim in all but the most egregious circumstances.")

38. See O'Connor, 480 U.S. at 726.

39. See William Stuntz, *Implicit Bargains, Government Power, and the Fourth Amendment,* 44 STAN. L. REV. 553, 579 (1992).

40. Smyth v. Pillsbury Corp., 914 F. Supp. 97, 100–01 (E.D. Pa. 1996).

41. *The Sociology of Georg Simmel,* Kurt H. Wolf, ed. (Glencoe, Ill.: Free Press, 1950), p. 354.

42. See Weise, *supra* note 2.

43. Strauss v. Microsoft Corp., 1995 U.S. Dist. LEXIS 7433, *10–*11 (S.D.N.Y. 1995).

44. Strauss v. Microsoft Corp., 856 F. Supp. 821, 825 (S.D.N.Y. 1994); Strauss v. Microsoft Corp., 814 F. Supp. 1186, 1194 (S.D.N.Y. 1993), cited in Strauss v. Microsoft Corp., 1995 U.S. Dist. LEXIS at *11.

45. See, e.g., Anne L. Lehman, *E-Mail in the Workplace: Question of Privacy, Property, or Principle?,* 5 COMMLAW CONSPECTUS 99, 99–100 & n. 7 (1997) (quoting Hal Coxson, a management lawyer, who notes that because employees can gain access to sexually explicit material on the Internet, which may create a hostile work environment, computer monitoring is justified to reduce an employer's liability for the potentially harassing acts of its employees).

46. See Hao-Nhien Vu, *Preventing Internet-Based Sexual Harassment in the Workplace,* L.A. DAILY J., October 3, 1997, at 5. See also Cheryl Blackwell Bryson & Michelle Day, *Workplace Surveillance Poses Legal, Ethical Issues,* NAT'L L.J., January 11, 1999, at B8 (warning that lawsuits based on offensive

e-mail "May [be] more easily resolved [if] the company monitor[s] its e-mail system, remove[s] 'offensive' material and discipline[s] employees for circulating these jokes"). For these and other examples of management lawyers advising employers to monitor their employees in order to avoid liability for sexual harassment, see generally Eugene Volokh, *Freedom of Speech, Cyberspace, Harassment Law, and the Clinton Administration* (1999) (unpublished draft on file with the author).

47. See Mitch Betts and Joseph Maglitta, "IS Policies Target E-mail Harassment," *Computerworld,* February 13, 1995, p. 12, cited in Jarrod J. White, *E-Mail @ Work.com: Employer Monitoring of Employee E-mail,* 48 ALA. L. REV. 1079, 1080 & n. 4 (1997).

48. Janice C. Sipior and Burke T. Ward, "The Dark Side of Employee Email," *Communications of the Association for Computing Machinery,* July 1, 1999, p. 88.

49. See, e.g., Dernovich v. City of Great Falls, Mont. Hum. Rts. Comm'n No. 9401006004 (November 28, 1995) (finding hostile environment based on off-color jokes and cartoons displayed in the workplace none of which were directed to complainant, and none of which were sexist or misogynistic, on the grounds that the jokes "ha[d] no humorous value to a reasonable person," and "offended [complainant] as woman").

50. See Baum, *supra* note 6, at 1038–40 & n. 139 (1997).

51. Ibid., p. 1039 n. 141.

52. Meritor Savings Bank, FSB v. Vinson, 477 U.S. 57, 73 (1986).

53. Barbara Lindemann and David Kadue, *Sexual Harassment in Employment Law* (Washington, D.C.: Bureau for National Affairs, 1992) (Appendix 6—"Sample Antiharassment Policy").

54. Ibid.

55. Rita Risser, *Fair Measures Management Law Consulting*

Group Research Report on the New Law of Sexual Harassment (1997) (excerpt from this report is available, with updates through 1999, at <http://fairmeasures.com/print/report/index2.html>).

56. Ibid.

57. Sipior and Ward, *supra* note 48.

58. Burlington Industries v. Ellerth, 118 S. Ct. 2257, 2270 (1998). See also Faragher v. City of Boca Raton, 524 U.S. 775, 118 S. Ct. 2275, 2293 (1998).

59. Ellerth, 118 S.Ct. at 2273 (Thomas, J., dissenting).

60. Mainstream Loudoun v. Board of Trustees of the Loudoun County Library, 2 F. Supp. 2d 783, 787 (E.D. Va. 1998) (citing Loudoun County Public Library, Policy on Internet Sexual Harassment (1997)).

61. Liane Hansen, *Loudoun County Internet,* National Public Radio Weekend Edition, November 2, 1997.

62. Complaint ¶¶92, 95, 127–129, cited in Mainstream Loudoun, 2 F. Supp.2d at 787.

63. Mainstream Loudoun v. Board of Trustees of the Loudoun County Library, 24 F. Supp.2d 552, 565 (E.D. Va. 1998). Case available at <http://www.eff.org/pub/Legal/Cases/Loudoun_library/HTML/19981123_opinion_order.html>.

64. Ibid., p. 567.

65. Jill Gerhardt-Powals and Matthew H. Powals, *Downloading Liability: Employers Could Face Harassment Claims Arising from Internet Use,* N.J. L.J., September 1, 1997, at 33.

66. Urofsky v. Allen, 995 F. Supp. 634, 643 (1998) (quoting Rankin v. McPherson, 483 U.S. 378, 384, (1987)) (alterations in original).

67. See, e.g., Loving v. Boren, 956 F. Supp. 953 (1997) (upholding University of Oklahoma's Internet filtering policy).

68. Davis v. Monroe County Board of Education, 119 S. Ct. 1661, 1666 (1999).

69. Ibid., pp. 1689–90 (Kennedy, J., dissenting).

70. Ibid., p. 1674 ("A university might not," Justice O'Connor wrote, "be expected to exercise the same degree of control over its students that a grade school would enjoy . . . and it would be entirely reasonable for a school to refrain from a form of disciplinary action that would expose it to constitutional or statutory claims").

71. See *Investigative report, Kavanagh v. Goddard College,* charge no. PA99-002, at 4–5 (March 10, 1999), quoted in Volokh, *supra* note 46.

72. Sipior and Ward, *supra* note 48.

CHAPTER 3: JURISPRURIENCE

1. Celia Farber, "The Trial," *Salon,* June, 1997, available at <http://www.salonmagazine.com/june97/spin4970609.html>.

2. Ibid.

3. Bonner v. Guccione, 1997 U.S. Dist. LEXIS 11382, *6-*10, 79 Fair Empl. Prac. Cas. (BNA) 1184, (S.D.N.Y. 1997), *aff'd in part, rev'd in part,* 178 F.3d 581 (2d Cir. 1999).

4. 42 U.S.C. § 2000e-2(a)(1) (1994).

5. As he introduced his amendment, Smith read a letter from a woman who noted an imbalance in the male and female population. "This is a grave injustice to womankind and something the Congress and President Johnson should take immediate steps to correct," the writer lamented. "Up until now, instead of assisting these poor unfortunate females in obtaining their 'right' to happiness, the Government has on several occasions

engaged in wars which killed off a large number of eligible males, creating an 'imbalance' in our male and female population that was even worse than before." 110 CONG. REC. 2577 (1964).

6. Michael Isikoff, *Uncovering Clinton: A Reporter's Story* (New York: Crown, 1999), p. 53 and n. 6.

7. Emily Listfield, "I Married a Billionaire: Profiles of Wealthy Wives," *Ladies' Home Journal,* September 1998, p. 196.

8. DeCintio v. Westchester County Medical Center, 807 F.2d 304, 307–08 (2d Cir. 1986).

9. James Lardner et al., "Cupid's Cubicles," *U.S. News & World Report,* December 14, 1998, p. 44.

10. "Love, Sex and the Bottom Line," *Success,* April 7, 1986, pp. 34–35.

11. See King v. Palmer, 598 F. Supp. 65, 69 (D.D.C. 1984), *rev'd,* 778 F.2d 878 (D.C. Cir. 1985).

12. King v. Palmer, 778 F.2d 878, 882 (D.C. Cir. 1985).

13. *EEOC Policy Guidance on Employer Liability for Sexual Favoritism,* reprinted in Lindemann and Kadue, *Sexual Harassment in Employment Law,* p. 656.

14. The policy then offered the example of a Charging Party (CP) who "alleges that she lost a promotion for which she was qualified because the co-worker who obtained the promotion" engaged in a consensual romantic affair with the supervisor. So long as the supervisor did not subject the coworkers to widespread and unwelcome sexual advances, the EEOC concluded, there would be no violation of Title VII, "because men and women were equally disadvantaged by the supervisor's conduct for reasons other than their genders." Ibid., p. 660.

15. Ibid., p. 657.

16. Ibid., p. 658.

17. Richard Posner, *An Affair of State* (Cambridge, Mass.: Harvard University Press, 1999), p. 138.

18. Lin Farley, *Sexual Shakdown: The Sexual Harassment of Women on the Job* (New York: McGraw-Hill, 1978), p. 15.

19. Catharine MacKinnon, *Sexual Harassment of Working Women* (New Haven: Yale University Press, 1979).

20. Ibid., p. 174 (emphasis omitted).

21. Ibid., p. 1.

22. Ibid., p. 182.

23. MacKinnon, *Toward a Feminist Theory of the State*, p. 127.

24. Ibid., pp. 127–28.

25. MacKinnon, *Sexual Harassment of Working Women*, pp. 218–21.

26. *In Harm's Way: The Pornography Civil Rights Hearings*, Catharine A. MacKinnon and Andrea Dworkin, eds. (Cambridge, Mass.: Harvard University Press, 1997), p. 41.

27. Ibid., p. 257.

28. Ibid., p. 270.

29. Ibid., p. 428.

30. American Booksellers Ass'n v. Hudnut, 771 F.2d 323, 328 (7th Cir. 1985), *aff'd* 475 U.S. 1001, 1065 S.Ct. 1172, 89 L.Ed.2d 291 (1986).

31. Ibid., p. 325 (citations omitted).

32. 29 C.F.R § 1604.11(a), reprinted in LINDEMANN, *supra* note 13, at 655.

33. Austen v. Hawaii, 759 F. Supp. 612, 683 (D. Haw. 1991), *aff'd* 967 F.2d 583 (9th Cir. 1992) (citing Ellison v. Brady, 924 F.2d 872 (9th Cir. 1991)).

34. See Anita Bernstein, *Treating Sexual Harassment With Respect*, 111 HARV. L. REV. 445, 465–67 & n. 122 (1997).

35. See Mark McLaughlin Hager, *Harassment As a Tort: Why Title VII Hostile Environment Liability Should Be Curtailed*, 30 CONN. L. REV. 375, 380 & n. 8 (citing Fortune 500 study finding 27 percent of sexual harassment complaints stemmed from intra-office relationships gone bad).

36. Lindemann and Kadue, *Sexual Harassment in Employment Law*, p. 135.

37. 510 U.S. at 25 (Ginsburg, J., concurring). See also Rebecca Hanner White, *There's Nothing Special About Sex: The Supreme Court Mainstreams Sexual Harassment*, 7 WM. & MARY BILL OF RTS. J. 725 (1999).

38. See, e.g., Robinson v. Jacksonville Shipyards, 760 F. Supp. 1486, 1502–06 (M.D. Fla. 1991).

39. Vicki Schultz, *Reconceptualizing Sexual Harassment*, 107 YALE L.J. 1683, 1716 n. 149 (1998).

40. Posner, *An Affair of State*, p. 36.

41. See Bernstein, *Treating Sexual Harassment With Respect*, *supra* note 34. See also Rosa Ehrenreich, *Dignity and Discrimination: Toward a Pluralistic Understanding of Workplace Harassment*, 88 GEORGETOWN L.J. 1 (1999).

42. Goffman, *Behavior in Public Places*, p. 14.

43. Lipsett v. University of Puerto Rico, 864 F.2d 881, 898 (1st Cir. 1988).

44. Lindemann, *Sexual Harassment in Employment Law*, p. 137 (citing Kouri v. Liberian Services, 55 Fair Empl. Prac. Cas. 124, 129 (E.D. Va. 1991) (complainant was bothered by supervisor's friendly notes, practice of escorting her to the bathroom and to her car, and visits to her at home and in the hospital; although she devised subtle schemes to let her supervisor know that she was happily married, each scheme "was a hopelessly indirect action that delivered an attenuated message")).

45. Highlander v. KFC National Management Co., 805 F.2d 644, 649 (6th Cir. 1986).

46. See, e.g., Durham v. Philippou, 968 F. Supp. 648, 659 (1997) (citing American Road Service Co. v. Inmon, 394 So. 2d 361, 365 (Ala. 1980)).

47. In Busby v. Truswal Systems Corp., 551 So. 2d 322, 324 (Ala. 1989), the Alabama Supreme Court held that a plant supervisor's extreme sexual harassment of his female employees created a jury question as to liability for the tort of outrage, as well as intrusion on seclusion. In Busby, the court found that there was evidence of at least seventeen incidents of harassment, including the supervisor's touching the plaintiffs and trying to follow one of them into the restroom, and that taken together, the incidents could constitute extreme and outrageous conduct.

48. RESTATEMENT (SECOND) OF TORTS § 652B (1977).

49. Edward J. Bloustein, "Privacy as an Aspect of Human Dignity: An Answer to Dean Prosser," in *Philosophical Dimensions of Privacy: An Anthology*, Ferdinand David Schoeman, ed. (Cambridge: Cambridge University Press, 1982), p. 165.

50. Ibid., p. 188.

51. Compare ibid. at 164 ("many of the cases allowing recovery for an intrusion expressly hold that special damages are not required . . . even in one of the rare cases in which serious mental distress was alleged, the court expressly says that recovery would be available without such an allegation") with Harris v. Forklift Systems, Inc., 510 U.S. 17, 21–23 (1993) (psychological injury not required to prove hostile environment harassment).

52. See, e.g., Cornhill Insurance PLC v. Valsamis, Inc., 106 F.3d 80, 85 (5th Cir. 1997) (Offensive comments and inappropriate

advances toward a plaintiff not cognizable as invasion of privacy under Texas law).

53. W. Page Keeton et al., PROSSER AND KEETON ON THE LAW OF TORTS § 117 (Supp. 1988).

54. See Phillips v. Smalley Maintenance Servs, 435 So. 2d 705, 708–09 (Ala. 1983).

55. McIsaac v. WZEW-FM Corp., 495 So. 2d 649, 651 (Ala. 1986).

56. See, e.g., Kelley v. Worley, 29 F. Supp. 2d 1304, 1311–12 (M.D. Ala. 1999) (supervisor's alleged conduct in sexually harassing female employee over her seven weeks of employment, including putting his hands under employee's dress and making other inappropriate physical contact, was sufficient to support invasion of privacy claim); see also Atmore Community Hospital v. Hayes, 719 So.2d 1190 (Ala. 1998) (Hospital employee presented substantial evidence that her supervisor committed invasion of privacy, where employee indicated that supervisor made several lewd comments, touched her legs, rubbed against her, asked her to meet him outside of work hours for other than business purposes, indicated that supervisor looked up her skirt on more than one occasion, but liability did not extend to employer); Cunningham v. Dabbs, 703 So. 2d 979, 982–83 (Ala. 1997) (genuine issue of material fact existed as to whether employer's sexual propositions and inappropriate physical contact with the employee unreasonably intruded into the employee's private affairs; plaintiff alleged that the individual defendant stuck his tongue in her ear, frequently rubbed her shoulders and repeatedly made lewd and suggestive comments to her); Kelley v. Troy State Univ., 923 F. Supp. 1494, 1503 (M.D. Ala. 1996) (denying motion to dismiss invasion of privacy claim where defendant: made sexually ex-

plicit remarks and jokes at the expense of women; made degrading personal comments to plaintiff, such as telling her she was having "a blond attack" when she made an error, and to "show a little leg" when a male colleague was coming to the office; and on several occasions, defendant struck plaintiff, either with an open hand or a closed fist).

57. McIsaac, 495 So. 2d at 650–51 (asking a co-employee for a date and making sexual propositions usually do not constitute an invasion of privacy); see also Culberson v. Philippou, 968 F. Supp. 648, 653–55 (M.D. Ala. 1997) (pattern of persistent propositions not actionable); Brassfield v. McClendon Furniture, Inc., 953 F. Supp. 1438, 1455–57 (M.D. Ala. 1996) (comments about the plaintiff's dress, her breasts, and a nude portrait which she brought to work, a request by the alleged harasser to sit in his lap and the alleged harasser's agreement with another employee who asked the plaintiff if she'd go skinny dipping did not constitute prying into the private matters of the plaintiff's life); Logan v. Sears, Roebuck & Co., 466 So. 2d 121, 123–24 (Ala. 1985) (defendant's statement that the plaintiff was as "queer as a three dollar bill" was not outrageous enough to maintain an invasion of privacy claim).

58. Hayes, 719 So. 2d at 1195 (relieving hospital employer of liability for tortious conduct of employee).

59. The precise legal standards for the intrusion on seclusion tort might be debated. For example, employers could be held liable for the invasive acts of their employees under a standard of negligence rather than actual knowledge—that is, if they should have known about the conduct, regardless of whether they actually knew about it. Similarly, employees could be held liable not only for intentional invasions of privacy but for negligent invasions.

60. See, e.g., Kelley, 923 F. Supp. at 1503 (M.D. Ala. 1996); see also Patterson v. Augat Wiring Systems, Inc., 944 F. Supp. 1509, 1523–24 (M.D. Ala. 1996) (Allegations of sexual harassment are sufficient to state claim for invasion of privacy in action by black female employee who alleged her privacy was invaded when her supervisor directed continual sexist, racial, and otherwise demeaning and profane statements toward her, allegedly telling her, while she was pregnant, that she should have her legs sewn together; employee also claimed that supervisor told her he was "so desperate for sex, he would even have it with black woman"). Note that these cases represent an exception to the majority rule that a person's privacy can't be invaded by oral statements, as opposed to those that are written or broadcast.

61. Goffman, *The Presentation of Self in Everyday Life,* p. 128.

62. Ibid.

63. Ibid., p. 130.

64. Barrington Moore, Jr., *Privacy: Studies in Social and Cultural History* (Armonk, N.Y.: M. E. Sharpe, 1984), pp. 44–45.

65. *The Sociology of Georg Simmel,* p. 321.

CHAPTER 4: PRIVACY IN COURT

1. Michelson v. U.S., 335 U.S. 469, 475–476 (1948).

2. Meritor Sav. Bank v. Vinson, 477 U.S. 57, 69 (1986).

3. Sanchez v. Zabihi, 166 F.R.D. 500, 502 (D.N.M. 1996).

4. Rex v. Smith, 11 Crim. App. 229 (1915).

5. See, e.g., Edward J. Imwinkeried, *Perspectives on Proposed Federal Rules of Evidence 413–415: Undertaking the Task of Reforming the American Character Evidence Prohibition: The Importance of Getting the Experiment Off on the Right Foot,*

22 Ford. Urb. L.J. 285, 297–298 (1995) (citing Thomas J. Reed, *Reading Gaol Revisited: Admission of Uncharged Misconduct in Sex Offender Cases,* 21 Am. J. Crim. L. 127 (1993)).

6. See David P. Bryden & Roger C. Park, *"Other Crimes" Evidence in Sex Offense Cases,* 78 Minn. L. Rev. 529, 572 (1994) (citing Allen J. Beck, Bureau of Justice Statistics, U.S. Dept. of Justice, *Recidivism of Prisoners Released in 1983* 6 (1989)).

7. See, e.g., Katharine K. Baker, *Once A Rapist? Motivational Evidence and Relevancy in Rape Law,* 110 Harv. L. Rev. 563 (1997) ("[T]he new rule will fail to reflect precisely what feminist scholarship of the past twenty-five years has established: the prevalence of rape in all social classes, among all races, and by all sorts of men . . . by singling out rape for special treatment, the new rule fosters the prevailing view that rape is different from other crimes because rapists are 'crazy' ").

8. Posner, *An Affair of State,* p. 215 and n. 36 (discussing Timur Kuran, *Ethnic Norms and Their Transformation Through Reputational Cascades,* 27 J. Legal Stud. 623 (1998); Eric A. Posner, *Symbols, Signals and Social Norms in Politics and the Law,* 27 J. Legal Stud. 765 (1998)).

9. Rochelle Gurstein, *The Repeal of Reticence* (New York: Hill & Wang, 1996), p. 47, citing E. L. Godkin, *Opinion-Moulding* (1869).

10. Posner, *An Affair of State,* p. 14.

11. Ibid., p. 81 and n. 36 (citing Amos Tversky and Daniel Kahneman, *Availability: A Heuristic for Judging Frequency and Probability,* 5 Cognitive Psychol. 207 (1973); Christine Jolls et al., *A Behavioral Approach to Law and Economics,* 50 Stan. L. Rev. 1471, 1477, 1518–1522 (1998); Timur Kuran &

Cass R. Sunstein, *Availability Cascades and Risk Regulation,* 51 STAN. L. REV. 769 (1999)).

12. Gurstein, *The Repeal of Reticence,* pp. 47–48.

13. John C. Danforth, *Resurrection: The Confirmation of Clarence Thomas* (New York: Viking, 1994), pp. 198–99.

14. For my own sins on this score, see Jeffrey Rosen, "Confirmations," *The New Republic,* December 19, 1994, p. 27.

15. *Nomination of Judge Clarence Thomas to be Associate Justice of the Supreme Court of the United States: Hearings Before the Senate Comm. on the Judiciary,* 102d Cong. 97 (1991).

16. McIsaac v. WZEW-FM Corp., 495 So. 2d 649, 651 (Ala. 1986) (asking an employee for dates and making sexual propositions do not constitute an invasion of privacy as a matter of law (citing Phillips v. Smalley Maintenance Serv. 435 So. 2d 705 (Ala. 1983)); see also Durham v. Philippou, 968 F. Supp. 648, 661 (M.D. Ala. 1997) (pattern of "persistent propositions" does not support an invasion of privacy claim); Brassfield v. Jack McLendon Furniture, Inc., 953 F. Supp. 1438 (M.D. Ala. 1996) (inquiring as to whether plaintiff wore panties and whether the initial "P" stood for "prostitute," in addition to commenting about a nude portrait which plaintiff brought to work and about plaintiff's breasts, did not constitute sufficient prying into the private aspects of plaintiff's life to sustain an invasion of privacy claim).

17. Rorie v. United Parcel Serv., 151 F.3d 757, 762 (8th Cir. 1998).

18. Jones v. Clinton, Plaintiff's Memorandum in Opposition to the Motion of Defendant Clinton to Limit Discovery at 5, Nov. 3, 1997, 990 F. Supp. 657 (E.D. Ark 1998) (No. LR-C-94-290).

19. Ibid., p. 20.

20. Jones v. Clinton, Order at 3, December 11, 1997.

21. Michael Issikoff, *Uncovering Clinton,* p. 357.

22. Jones v. Clinton, Memorandum in Support of Jane Doe #6's Motion for Protective Order and Motion to Quash at 8, January 20, 1998.

23. Johnson v. Wal-Mart Stores, Inc., 987 F. Supp. 1376, 1381, 1397 (M.D. Ala. 1997).

24. Robert C. Post, *Cultural Heterogeneity and Law: Pornography, Blasphemy, and the First Amendment,* 76 CALIF. L. REV. 297, 305 (1988).

25. Posner, *An Affair of State,* p. 209 (discussing Thomas Nagel, *Concealment and Exposure,* 27 PHIL. & PUB. AFF. 3 (1998)).

26. Ibid. ("[F]launting makes it a public issue").

CHAPTER 5: PRIVACY IN CYBERSPACE

1. Fox Butterfield, "Pornography Cited in Ouster of Harvard Divinity School Dean," *The New York Times,* May 20, 1999, p. A21.

2. See Michael Walzer, *Spheres of Justice* (New York: Basic Books, 1983), pp. 295–98.

3. "Richard Hemingway," "Porn, the Harvard Dean, and Tech Support," *Salon,* May 21, 1999, <http://www.salon.com./tech/feature/1999/05/21/tech_support/index1.html>.

4. Ibid.

5. See J. M. Lawrence, "Harvard Cyberporn Users Get a Shock," *The Boston Herald,* February 19, 1995, p. 1.

6. See Lawrence Lessig, *The Path of Cyberlaw,* 104 YALE L.J. 1743, 1748 (1995).

7. For computers directly on the Internet (continuous connections where you don't have to dial up), the IP address is a four-number list (128.53.100.5), where each number (between 0

and 256 in the original IP regime) designates a smaller subclass of the Internet. When you dial up via an ISP, you are given a dynamic IP address which changes each time you log in. In this case, it is some other key, like your e-mail address or password (along with the phone number you dial from), which links the dynamic IP address to your personal identity.

8. See Christian Baekkelund et al., *A Framework for Privacy Protection*, White Paper presented to MIT 6.805/STS085: Ethics and Law on the Electronic Frontier; and Harvard Law School: The Law of Cyberspace—Social Protocols (December 10, 1998) <http://cyber.law.harvard.edu/ltac98/privacy.html>.

9. Saul Hansell, "Big Web Sites to Track Steps of Their Users," *The New York Times,* August 16, 1998, p. A1.

10. See Jerry Berman and Deirdre Mulligan, *The Internet and the Law: Privacy in the Digital Age: Work in Progress,* 23 Nova L. Rev. 549, 559 (1999). See also Bob Tedeschi, "Critics Press Legal Assault on Tracking of Web Users," *The New York Times,* February 7, 2000, p. C1.

11. See A. Michael Froomkin, *Regulation and Computing and Information Technology: Flood Control on the Information Ocean: Living with Anonymity, Digital Cash, and Distributed Databases,* 15 J.L. & Com. 395, 482 (1996).

12. Jerry Kang, *Information Privacy in Cyberspace Transaction,* 50 Stan. L. Rev. 1193, 1198–99 (1998).

13. Ibid., pp. 1198–99.

14. Julie E. Cohen, *A Right to Read Anonymously: A Closer Look at "Copyright Management" in Cyberspace,* 28 Conn. L. Rev. 981, 1019 (1996).

15. Bill Carter, "Will This Machine Change Television?," *The New York Times,* July 5, 1999, p. C1.

16. See Cohen, *supra* note 14, at 1031 & n. 213.

17. Ulrika Ekman Ault, Note: *The FBI's Library Awareness Program: Is Big Brother Reading Over Your Shoulder?*, 65 N.Y.U.L. Rev 1532, 1535–36 (1990).

18. See Cohen, *supra* note 14, at 1032.

19. McIntyre v. Ohio Election Comm'n, 514 U.S. 334, 343 n. 6, (1995).

20. See ibid., 514 U.S. at 361 (Thomas, J., concurring) citing J. Alexander, *A Brief Narrative of the Case and Trial of John Peter Zenger*, S. Katz, ed. (Cambridge, Mass.: Harvard University Press, 1972), pp. 9–19, 22–23.

21. See Lamont v. Postmaster General, 381 U.S. 301, 307 (1965).

22. Stanley v. Georgia, 394 U.S. 557, 565 (1969).

23. See NAACP v. Alabama *ex rel.* Patterson, 357 U.S. 449, 466 (1958).

24. See *McIntyre*, 514 U.S. at 361–66, 115 S. Ct. at 1526–28 (Thomas, J., concurring) (describing attempts in the founding era to discover the identities of authors of anonymous pamphlets).

25. See Robert O'Harrow Jr., "Prescription Sales, Privacy Fears: CVS, Giant Share Customer Records with Drug Marketing Firm," *The Washington Post*, February 15, 1998, p. A1. In the face of public criticism, CVS and Giant Food agreed to stop the disclosures. See Robert O'Harrow Jr., "CVS Also Cuts Ties to Marketing Service; Like Giant, Firm Cites Privacy on Prescriptions," *The Washington Post*, February 19, 1998, p. E1.

26. See, e.g., Robert Pear, "Future Bleak for Bill to Keep Health Records Confidential," *The New York Times*, June 21, 1999, p. A12. At the end of 1999, after Congress failed to act, the Clinton administration proposed a series of medical privacy regulations.

27. See H.R. 1972, 105th Cong. (1997) (Children's Privacy Protection and Parental Empowerment Act).

28. See H.R. 2368, 105th Cong. (1997) (Data Privacy Act).

29. See H.R. 98, 105th Cong. (1997) (Consumer Internet Privacy Protection Act).

30. See HR 1748, 105th Cong. (1997) (Netizens Protection Act).

31. See S. 771, 105th Cong. (1997) (Unsolicited Commercial Electronic Mail Choice Act).

32. See Cohen, *supra* note 14, at 1037.

33. Even 128-bit encryption is breakable with current technology with months or years to spare, and thus can pierce someone's identity over the long term. Meanwhile, computers are still getting faster and cryptography is advancing as I write. A 512-bit code may be similarly vulnerable in a few years time. See Steve Gold, *Expert Goes Public on Encryption Cracking Engine,* Newsbytes (August 13, 1999) <http://www.newsbytes.com> (reporting on a new optical computer theoretically capable of breaking 512-bit encryption in days).

34. *McIntyre,* 514 U.S. at 384 (Scalia, J., dissenting).

35. The Georgia law made it a crime for "for any person . . . knowingly to transmit any data through a computer network . . . for the purpose of setting up, maintaining, operating, or exchanging data with an electronic mailbox, home page, or any other electronic information storage bank or point of access to electronic information if such data uses any individual name . . . to falsely identify the person . . ." and similarly for any person knowingly to transmit any data through a computer network "if such data uses any . . . trade name, registered trademark, logo, legal or official seal, or copyrighted symbol . . . which would falsely state or imply that such per-

son . . . has permission or is legally authorized to use [it] for such purpose when such permission or authorization has not been obtained." Ga. Code. An. 16-9-93.1 (Mitchie 1999, LEXIS through 1999 Sess.) See generally Donald J. Karl, *State Regulation of Anonymous Internet Use After* ACLU of Georgia v. Miller, 30 ARIZ. ST. L.J. 513, 517 & n. 43 (1998).

36. Georgia v. Miller, 977 F. Supp. 1228, 1232 (1997) (quoting *McIntyre,* 514 U.S. at 348).

37. Ibid., p. 1233.

38. See Buckley v. American Constitutional Law Foundation, Inc., 119 S. Ct. 636, 645 (1999).

39. For proposals to create property rights in personal information, see Peter P. Swire and Robert E. Litan, *None of Your Business: World Data Flows, Electronic Commerce, and the European Privacy Directive* (Washington, D.C.: Brookings Institute Press, 1998), pp. 86–87; see also Richard S. Murphy, *Property Rights in Personal Information: An Economic Defense of Privacy,* 84 GEO. L.J. 1831 (1996). For a critique of the property rights approach, see Jessica Litman, *Information Privacy/Information Property,* 52 STANFORD L. REV. (2000).

40. Ginsberg v. New York, 390 U.S. 629 (1968).

41. Martin Rimm, *Marketing Pornography on the Information Superhighway,* 83 GEO. L.J. 1849, 1867 (1995).

42. Philip Elmer-Dewitt and William Dowell, "Fire Storm on the Computer Nets: A New Study of Cyberporn, Reported in a Time Cover Story, Sparks Controversy," *Time,* July 24, 1995, p. 57.

43. Reno v. ACLU., 521 U.S. 844, 878–79 (1997).

44. In an effort to respond to O'Connor's suggestion, Congress passed a daughter of the CDA bill, which is called the Child Online Protection Act and is drafted a bit more carefully than

its predecessor. It narrows the category of proscribed material to speech that is posted on the Web and meets the legal definition of being obscene for minors; and it expands the list of defenses that are available to people who have tried, in good faith, to restrict access to minors. COPA, however, suffers from the same underlying problem as its predecessor: the current technologies of age verification are economically unfeasible for noncommercial Web sites. Credit cards can't be used unless they accompany a commercial transaction, and, anyway, not everyone has a credit card; digital certificates are still expensive for individual Web sites to accept and screen; and adult users may be deterred from entering controversial (but legally protected) sites if they have to identify themselves in advance. For this reason, COPA, like the CDA, was enjoined by a lower court soon after it was passed. ACLU v. Reno, 31 F.Supp.2d 473, 495 (E.D. Pa. 1999), cited in Timothy Zick, *Congress, The Internet, and the Intractable Pornography Problem: The Child Online Protection Act of 1998,* 32 CREIGHTON L. REV. 1147, 1188 & n. 270 (1999).

45. Lawrence Lessig, *Code and Other Laws of Cyberspace* (New York: Basic Books, 1999), p. 270.

46. Andrew L. Shapiro, *The Control Revolution: How the Internet Is Putting Individuals in Charge and Changing the World We Know* (New York: Public Affairs, 1999), p. 173.

47. See, e.g., ACLU v. Reno, 929 F. Supp. 824, 838 (E.D. Pa. 1996) *jurisdiction noted* 518 U.S. 1025, *aff'd* 521 U.S. 844 (1997).

48. Rohan Samarajiva, "Interactivity as Though Privacy Mattered," in *Technology and Privacy: The New Landscape,* Philip E. Agre and Marc Rotenberg, eds. (Cambridge, Mass.: MIT Press, 1997), p. 283.

49. Shapiro, *The Control Revolution,* p. 109.

50. See Patrick Garry, *Scrambling for Protection: The New Media and the First Amendment* (Pittsburgh: University of Pittsburgh Press, 1994), pp. 8–9.

51. Mike Godwin, *Cyber Rights: Defending Free Speech in the Digital Age* (New York: Times Books, 1998), pp. 91–92.

52. Ibid., pp. 88–89.

53. The relevant section says, "No provider or user of an interactive computer service shall be treated as the publisher or speaker of any information provided by another information content provider." 47 U.S.C. § 230(c)(1) (1999, LEXIS through 1999 Sess.). In fact, bookstores are liable for falsehoods in their books, so long as they know (or have reason to know) about the falsehoods. In this respect, the Communications Decency Act gives Internet Service providers more protection than bookstores have traditionally had.

54. Zeran v. America Online, Inc., 129 F.3d 327, 330–31 (4th Cir. 1997).

55. Blumenthal v. Drudge, 992 F. Supp. 44, 51–52 (1998).

56. Robert C. Post, *Symposium: Free Speech and Community: Community and the First Amendment*, 29 ARIZ. ST. L.J. 473, 476–79 (1997).

57. Shapiro, *The Control Revolution*, p. 141 (citing Drudge, *The Media Should Apologize*, text of address to the Wednesday Morning Club, September 10, 1998, <http://www.frontpagemag.com/archives/drudge/wmscpeech.htm>).

58. See, e.g., David K. McGraw, *Sexual Harassment in Cyberspace: The Problem of Unwelcome E-mail*, 21 RUTGERS COMPUTER & TECH. L.J. 491, 503–04 (1997).

59. See Volokh, *Freedom of Speech in Cyberspace from the Listener's Perspective: Private Speech Restrictions, Libel, State*

Action, Harassment, and Sex, 1996 U. CHI. LEGAL F. 377, 411 (citing *Rowan v. United States Post Office,* 397 U.S. 728, 90 S.Ct. 1484, 25 L.Ed.2d 736 (1970)).

60. *Letter from John E. Palomino, Regional Civil Rights Director for United States Department of Education, Office of Civil Rights, to Dr. Robert F. Agrella, President of Santa Rosa Junior College, in case no.* 09-93-2202, at 2 (June 23, 1994).

61. See the excellent account of the Santa Rosa case in Godwin, *Cyber Rights,* pp. 105–110.

62. Volokh, *supra* note 59, at 419–20.

63. Godwin, *Cyber Rights,* pp. 110–14.

64. Ibid. at 111. In a pyrrhic attempt at protest, the college's attorney added the proviso that nothing in the policy should be construed "as violating any person's rights of expression set forth in the Equal Access Act or the First Amendment to the United States Constitution." Ibid. at 116–17.

65. Johnson v. Sawyer, 47 F.3d 716, 731 (5th Cir. 1995), *later proceedings vac'd, remanded, and reassigned* 120 F.3d 1397 (5th Cir. 1997) (quoting *Industrial Found. of the South v. Texas Indus. Accident Bd.,* 540 S.W.2d 668, 683–85 (Tex. 1976) (citations omitted).

66. Godwin, *Cyber Rights,* p. 105.

EPILOGUE: WHAT IS PRIVACY GOOD FOR?

1. Mill, *On Liberty,* p. 7.

2. *The Sociology of George Simmel,* p. 404.

3. See Lawrence Lessig, *The Censorships of Television* (1999), p. 24, available at <http://cyber.law.harvard.edu/lessig.html>.

4. Ibid., p. 19.

5. Ferdinand Tönnies, *Community and Society,* Charles P. Loomis, ed. (East Lansing, Mich.: Michigan State Univ. Press, 1957), pp. 33, 64–65.

6. Ferdinand David Schoeman, *Privacy and Social Freedom* (New York: Cambridge University Press, 1992), pp. 149–50.

7. For the differences among the courts, see I. J. Schiffres, Annotation, *Invasion of Right of Privacy By Merely Oral Declarations,* 19 A.L.R. 1318 (1968 & Supp. 1999).

8. Schoeman, *Privacy and Social Freedom,* p. 149.

9. See, e.g., Howard Kurtz, "The 'Love Child' Story Turns Into an Orphan," *The Washington Post,* January 11, 1999, p. C1.

10. Michael Kinsley, "In Defense of Matt Drudge," *Time,* February 2, 1998, p. 41.

11. Edward Dorn, *Gunslinger 1 & 2* (London: Fulcrum Press, 1970), pp. 31–32.

12. Erving Goffman, *The Presentation of Self in Everyday Life,* p. 115.

13. Alan F. Westin, *Privacy and Freedom* (New York: Atheneum, 1967), p. 36.

14. Richard Posner, "An Economic Theory of Privacy," in *Philosophical Dimensions of Privacy,* pp. 333, 337–38 ("But each of us should be allowed to protect ourselves from disadvantageous transactions by ferreting out concealed facts about other individuals that are material to their implicit or explicit self-representations").

15. Richard A. Wasserstrom, "Privacy: Some Arguments and Assumptions," in *Philosophical Dimensions of Privacy,* pp. 317, 331–32.

16. Brin, *The Transparent Society,* p. 23.

17. See also Ferdinand Schoeman, "Privacy: Philosophical Dimensions of the Literature," in *Philosophical Dimensions of Pri-*

vacy, pp. 1, 30 ("Perhaps people can behave in different ways in different contexts without exhibiting inauthenticity in any of these contexts. People may really be complex in the sense that they are not basically one thing").

18. Edward A. Shils, *The Torment of Secrecy: The Background and Consequences of American Security Policies* (Glencoe, Ill.: Free Press, 1956), p. 155.

19. Ibid., p. 201.

20. Benjamin Ginsberg and Martin Shefter, *Politics by Other Means: The Declining Importance of Elections in America* (New York: Basic Books, 1990), p. 26.

21. See, e.g., S. Elizabeth Wilborn, *Revisiting the Public/Private Distinction: Employee Monitoring in the Workplace,* 32 GA. L. REV. 825, 837 & n. 47 ("[T]wice as many electronically monitored workers reported wrist pains and 20% more reported neck pains, as compared with those who were not monitored, and . . . monitored employees noted higher incidents of depression, tension, anger, and extreme anxiety" (citing Michael J. Smith et al., University of Wis.-Madison Dep't of Engineering, *Electronic Performance Monitoring and Job Stress in Telecommunications Jobs* 1, 5, 20 (1990)).

22. Stanley I. Benn, "Privacy, Freedom and Respect for Persons," in *Philosophical Dimensions of Privacy,* pp. 223, 229–30.

23. Posner, *supra* note 14, at 339.

24. Michel Foucault, *Discipline and Punish: The Birth of the Prison,* Alan Sheridan, trans. (New York: Vintage Books, 1979), p. 201.

25. Charles Fried, "Privacy," in *Philosophical Dimensions of Privacy,* pp. 203, 205.

26. Edith Wharton, "The Other Two," in *Roman Fever and Other Stories* (New York: Scribner Paperback Fiction, 1997), p. 81.

27. Ralph Waldo Emerson, "Friendship," in *Selected Essays* (Chicago: Peoples Book Club, 1949), p. 138.

28. Westin, *Privacy and Freedom,* p. 34 (citing Leontine Young, *Life Among the Giants* [New York: McGraw-Hill, 1966]).

29. Ibid., p. 37. Westin notes that Yale studies of "brainstorming" in 1958 found that work groups produced fewer ideas than periods of individual effort by the same number of people.

30. Virginia Woolf, *A Room of One's Own* (New York: Harcourt Brace, 1989), p. 182.

31. Marcel Mauss, "A Category of the Human Mind: The Notion of Person; The Notion of Self," in *The Category of Person: Anthropology, Philosophy, History,* Michael Carrithers et al., eds. (New York: Cambridge University Press, 1985), p. 14.

32. Ibid., p. 22.

33. Roger Fulford, *The Trial of Queen Caroline* (New York: Stein & Day, 1968), p. 120.

Index

ALL THE LAWS BUT ONE
Civil Liberties in Wartime
by William H. Rehnquist

Abraham Lincoln suspended the writ of *habeas corpus* early in the Civil War and later imposed limits on freedom of speech. During World War II, the government forced 100,000 U.S. residents of Japanese descent into detainment camps. Through these and other incidents, William H. Rehnquist examines the balance between the national interest and personal freedoms and elucidates how the Supreme Court has interpreted the Constitution in the past—and draws guidelines for how it should do so in the future.

History/Law/0-679-76732-0

OUT OF ORDER
by Thomas E. Patterson

According to Thomas E. Patterson, the road to the presidency has led through the newsrooms, which in turn have imposed their own values on American politics. This is why Bill Clinton's draft record was deemed more newsworthy than his policy statements, and why George Bush's masculinity and Ronald Reagan's theatrics with a microphone profoundly affected their presidential campaigns. *Out of Order* presents a devastating inquest into the press's hijacking of the campaign process—and shows what we can do to win it back.

Current Affairs/0-679-75510-1

BREAKING THE NEWS
How the Media Undermine American Democracy
by James Fallows

Why do Americans mistrust the news media? Is it because programs such as *The McLaughlin Group* reduce participating journalists to so many shouting heads? Or because, increasingly, the profession treats issues as complex as health-care reform and foreign policy as exercises in political gamesmanship? Drawing on his own experience as a journalist—and on the gaffes of colleagues from George Will to Cokie Roberts—James Fallows shows why the media have not only lost our respect but also alienated us from our public life.

Current Affairs/0-679-75856-9